SPEED READ
FERRARI

Brimming with creative inspiration, how-to projects, and useful information to enrich your everyday life, Quarto Knows is a favorite destination for those pursuing their interests and passions. Visit our site and dig deeper with our books into your area of interest: Quarto Creates, Quarto Cooks, Quarto Homes, Quarto Lives, Quarto Drives, Quarto Explores, Quarto Gifts, or Quarto Kids.

First Published in 2018 by Motorbooks, an imprint of The Quarto Group, 100 Cummings Center, Suite 265-D, Beverly, MA 01915, USA. T (978) 282-9590 F (978) 283-2742 QuartoKnows.com

Motorbooks titles are also available at discount for retail, wholesale, promotional, and bulk purchase. For details, contact the Special Sales Manager by email at specialsales@quarto.com or by mail at The Quarto Group, Attn: Special Sales Manager, 100 Cummings Center, Suite 265-D, Beverly, MA 01915, USA.

10 9 8 7 6 5 4 3 2

ISBN: 978-0-7603-6040-8

Library of Congress Cataloging-in-Publication Data
Names: Lerner, Preston, 1956- author.
Title: Speed read Ferrari : the history, technology and design behind Italy's
   legendary automaker / Preston Lerner.
Description: Minneapolis, MN : Motorbooks, 2018.
Identifiers: LCCN 2017054064 | ISBN 9780760360408 (paperback)
Subjects: LCSH: Ferrari automobile--History. | BISAC: TRANSPORTATION /
   Automotive / History. | TRANSPORTATION / Automotive / Pictorial. |
   TECHNOLOGY & ENGINEERING / Automotive.
Classification: LCC TL215.F47 L47 2018 | DDC 629.2220945--dc23
LC record available at https://lccn.loc.gov/2017054064

Acquiring Editor: Darwin Holmstrom
Project Manager: Alyssa Bluhm
Series Creative Director: Laura Drew
Cover and interior illustrations by Chris Rathbone

Printed in China

SPEED READ
FERRARI

THE HISTORY, TECHNOLOGY AND DESIGN
BEHIND ITALY'S LEGENDARY AUTOMAKER

PRESTON LERNER

INTRODUCTION
WHAT'S IN A NAME

Ferrari. No car comes with more cachet. None boasts a more storied history.

Ferrari. Three syllables that roll off the tongue with ineffable Italian panache. The name evokes vivid images of sleek sports cars and shark-like racers. Oh, and the sounds! The click of the fuel pump, the satisfying *thunk* of the shift lever finding a gear in the gated shifter, the intoxicating aria sung by a twelve-cylinder engine approaching redline.

Ferrari. The cars have been driven by celebrities and beautiful people ranging from sheiks and princes to the most exalted motorsports royalty. The *belle macchine* from Maranello carried them to fame, fortune, and, on occasion, death.

But always—always!—in Ferrari's inimitable style.

From the start, the company was the corporate embodiment of its imperious founder, Enzo Ferrari. A former race car driver turned impresario of Alfa Romeo's works racing team, Ferrari never aspired to be a captain of industry. No, Ferrari created his eponymous marque for the sole purpose of winning races, and he built and sold street cars only to support his passion for motorsports. This freed him to adopt an uncompromising approach to car design and a cavalier attitude toward his customers. The products of his single-mindedness—some called it obstinacy—were race cars that triumphed everywhere from the Grand Prix of Monaco to the 24 Hours of Le Mans and road cars that set standards for elegance, performance, and exclusivity.

In 1969, Ferrari became part of the vast Fiat conglomerate and entered the modern industrial age. Once built by hand by the dozen in a grimy, cramped factory, cars are now stamped out by the thousands in a state-of-the-art assembly plant filled with robots rather than artisans. And yet, despite all the changes, every car emblazoned with Ferrari's famed prancing horse is sprinkled with some of the same fairy dust found in the 166 MM Barchetta, the 250 GTO, and the FS70H that's still winning Formula 1 races to this day.

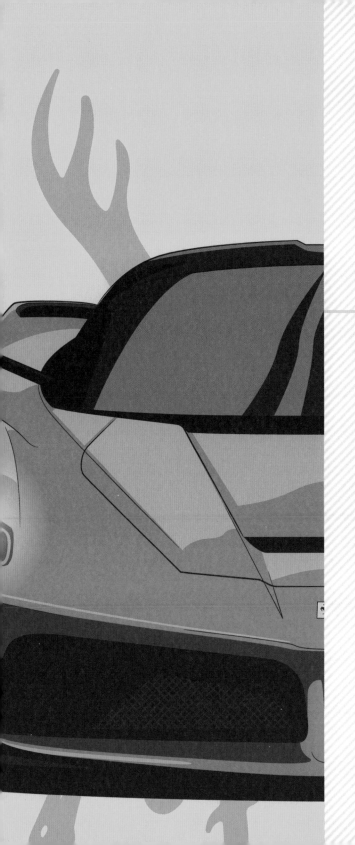

A LEGEND IS BORN

A LEGEND IS BORN
ENZO ASCENDANT: THE PREWAR YEARS

FUN FACT

Ferrari's greatest victory as a driver, in the Coppa Acerbo in 1924, earned him a seat in an all-conquering Alfa Romeo P2 Grand Prix car. But he fell ill—or, some say, lost his confidence—before he could drive it, and he raced only intermittently thereafter.

HISTORICAL TIDBIT

During World War II, Italian officials ordered Ferrari to move his factory out of Modena in the hope of safeguarding it. Ironically, his new factory in Maranello was pummeled by three Allied bombing raids.

KEY PERSON

Ferrari was instrumental in enticing legendary engineers Luigi Bazzi and Vittorio Jano from Fiat to Alfa Romeo in the 1920s. After the war, both men worked for Ferrari in Maranello.

Few car companies embody the character of their founder more faithfully than Ferrari. This is a marque with a famous split personality, obsessed by street cars at one end of the spectrum and race cars at the other. So, too, was Enzo Ferrari himself riven by contradictions—crude yet courtly, callous yet affectionate, quick-thinking yet pig-headed, narrow-minded yet larger than life. He was born in 1898 in the northern Italian town of Modena, the son of the owner of a small machine shop. After World War I, he moved to Turin and found work at a company that stripped war-surplus trucks down to bare chassis. He immediately started racing, with modest success, and soon finagled a position at Alfa Romeo. Ferrari quickly made himself indispensable to the burgeoning company, selling cars and poaching engineers. Although he wasn't a first-class driver, his racing experience helped him persuade several wealthy enthusiasts to bankroll the formation in 1929 of a private race team, Società Anonima Scuderia Ferrari. In 1933, Scuderia Ferrari became Alfa Romeo's works operation, with Ferrari presiding over a race program that featured luminaries such as Tazio Nuvolari and Achille Varzi. Relations with Alfa Romeo soured during the late 1930s, as Mercedes-Benz and Auto Union dominated Grand Prix racing. Ferrari returned to Modena to form a machine-tool company he named Auto Avio Costruzione. But he couldn't resist the lure of racing, and he commissioned a pair of Fiat-based entries for the Mille Miglia in 1940. Both of the so-called 815s led their class. Neither finished. Then came World War II and the end of the prologue of the Ferrari story.

A LEGEND IS BORN
STAKING A CLAIM: 1947 TO 1969

FUN FACT

The first car to bear the Ferrari nameplate, the 125 S, was what Enzo called "a promising failure" when it debuted at Piacenza in 1947. But it won its next race, in Rome, two weeks later.

HISTORICAL TIDBIT

The 250 GT, introduced in 1954, is generally considered to be the first genuine "production" Ferrari. The basic chassis, which was used for the next decade, worked equally well on road and track.

KEY PERSON

In 1963, Enzo and Henry Ford II hammered out, but ultimately refused to sign, a contract that would have created two new companies—Ford-Ferrari to build and sell sports and grand touring cars and Ferrari-Ford to go racing.

It's tempting to think of the 1950s as Ferrari's golden years because so many immensely valuable cars date from this era. But these cars are so collectible largely because so few of them were built. Ferrari painstakingly assembled running chassis by hand, while the bodies were designed and built elsewhere. It wasn't until 1958 that the factory got its first assembly line (to use the term charitably). In 1960, a total of 306 cars were built in Maranello; GM knocked out more than that in an hour. But if Ferrari got off to a slow start on the production side of the business, it sprinted out of the box on the racing front. A Ferrari 166 MM won the first 24 Hours of Le Mans staged after World War II, and Ferrari 500 single-seaters swept every World Championship race in 1952. By the 1960s, though, as racing costs skyrocketed, it became clear that profits from production cars could no longer finance the entire motorsports program. Enzo spent weeks negotiating the sale of his company to the Ford Motor Company before abruptly calling things off when he realized that he'd no longer be the despot-in-chief. This, after all, was a man who'd fired, en masse, the directors of engineering, sales, purchasing, personnel, and the race team, in 1961 for a perceived lack of loyalty—an episode known as "The Purge." But Ferrari had an ace in the hole in the form of his ties to Gianni Agnelli, the tycoon in charge of the Fiat empire. In 1969, Fiat bought a 40-percent stake in Ferrari (with an option for an additional 50 percent when Enzo died), and the marque's future was assured.

A LEGEND IS BORN
THE FIAT CONNECTION: 1969 TO 1988

FUN FACT

The most famous Ferrari of the 1980s wasn't a Ferrari or from the 1980s. It was a modern replica of the 1961 250 GT California Spider that came to a memorably bad end in *Ferris Bueller's Day Off*.

HISTORICAL TIDBIT

A month after Enzo's death, at the Italian Grand Prix at Monza, race leader Ayrton Senna collided with a backmarker with two laps to go, gifting Ferrari with a miraculous 1–2 finish. It was the only race McLaren would lose all year.

KEY PERSON

Ferrari planned to race at the Indianapolis 500 in 1986. Formula 1 designer Gustav Brunner directed the construction of the chassis. But technical director John Barnard nixed the program—dubbed Project 637—to focus on F1.

The 1970s were memorable for a lot of reasons—*Star Wars*, the Walkman, the first Earth Day celebration—but cars generally weren't one of them. Two oil crises and new safety regulations confounded automakers, especially those specializing in sexy, high-performance vehicles. Ferrari benefitted from the corporate backing, technical know-how, and physical resources of Fiat, and production skyrocketed from 928 in 1970 to 2,221 by the end of the decade. But for every electrifying Berlinetta Boxer, there was an uninspired 400 Automatic, and critics whispered that the company was pumping up sales by pumping out high-priced Fiats in Ferrari clothing. The news was better on the motorsports side. Even as purists howled about the unsentimental decision to scrap the prototype sports car program, the extra emphasis placed on Formula 1 produced four constructors' titles and two second-place finishes in six years, with Niki Lauda and Jody Scheckter claiming three championships between them. Still, the coming of a new decade brought more hard times. In 1982, the only Ferraris being sold in America were the unloved 400i, the underwhelming Mondial, and the long-in-the-tooth 308. (The BBi wasn't imported because it couldn't meet US safety or emissions standards.) Then, the F1 team was hit with two gut punches—the death of Gilles Villeneuve and Didier Pironi's career-ending crash. During the fallow years that followed, British engineers were granted control of the F1 program, and the design office moved to England. Enzo, by then old and feeble, summoned the strength to greenlight one last car, the magnificent F40, before his death in 1988. The marque he founded would thrive after he died. But its character would never be the same.

A LEGEND IS BORN
FERRARI AFTER ENZO: 1988 TO 2016

FUN FACT

Ferrari owns a huge, state-of-the-art factory in Maranello, a smaller one in Modena, and a race shop/office at its test track in Fiorano. The company also owns the Mugello Circuit in Tuscany, though not, ironically, the Autodromo Internazionale Enzo e Dino Ferrari in Imola.

HISTORICAL TIDBIT

For buyers who think Ferraris aren't exclusive enough, the company created its Tailor Made program to personalize production cars with custom paint, materials, features, and pretty much anything else you can think of (and pay for).

KEY PERSON

Although Montezemolo showed promise as a young rally driver for Lancia, he came to Enzo Ferrari's attention through his cheeky comments about racing on a talk radio show.

Nature abhors a vacuum. Without a clear line of succession, Ferrari floundered after Enzo's death in 1988. The Formula 1 team was in disarray, and the 348, which was supposed to lead the way to the digital future, was short-listed for the Worst Ferrari Ever award. Sales plummeted from 4,001 the year Enzo died to 2,345 in 1993. Fiat had exercised its option to buy 50 percent of the company and now owned all but the 10 percent stake held by Enzo's son Piero. So, Fiat chief Gianni Agnelli gave a protégé, Luca di Montezemolo, carte blanche to restore Ferrari to its rightful place. Montezemolo had already done a brief stint at Ferrari, overseeing the resurrection of the moribund Formula 1 program in the mid-1970s, before leaving to run Cinzano and organize the 1990 World Cup. As the unhappy owner of a 348, he understood that Ferrari had to upgrade its meat-and-potatoes products, and the thoroughly modern F355 was the result. Before long, Ferrari was in danger of selling *too many* cars, prompting Montezemolo to cap production at about seven thousand units a year to maintain exclusivity. Meanwhile, Michael Schumacher won five consecutive F1 World Championships and spread the Ferrari gospel to *tifosi* all over the world. Eventually, Ferrari became more profitable than its parent company, Fiat Chrysler Automobiles. But even as plans for a corporate separation were being floated, Montezemolo was butting heads with FCA chief Sergio Marchionne, who was the yin to his yang. Agnelli had died in 2003. With no one to protect him, Montezemolo was unceremoniously kicked to the curb in 2014. Two years later, when Ferrari was officially spun off from FCA, Marchionne became its chairman, CEO, and executive director.

A LEGEND IS BORN
ALONE AGAIN: FERRARI CIRCA 2017

In 1969, Ferrari needed the backing of Fiat to remain afloat. But by 2016, Ferrari's bottom line was so buoyant that a new firm, dubbed Ferrari N.V., was spun off from Fiat Chrysler Automobiles, which had become a financial anchor. In 2016, Ferrari generated $449 million in net profits on revenues of $3.7 billion, and as the marque celebrated its seventieth anniversary in 2017, it planned to build about eight thousand cars the following year. Meanwhile, the company was flush enough to spend nearly half a billion dollars annually on its Formula 1 program. More than a mere automobile manufacturer, Ferrari is also a globally recognized luxury symbol, and few companies have exploited their brand image more relentlessly. In addition to forty-five authorized Ferrari Stores from Milan to Miami, there are also two theme parks—Ferrari World in Abu Dhabi and Ferrari Land near Barcelona. Ferrari has even managed to monetize its history through its Classiche Department, which works on about forty vintage cars and completes roughly ten full restorations a year. For a substantial fee, cars built more than twenty years ago are eligible for Classiche "certificates of authenticity," which have become must-have documents for sellers hoping to get top dollar for their cars. While new cars remain the core of Ferrari's business, the company that started with a single model in 1947 now builds two-seaters, four-seaters, coupes, spiders, ultra-limited-edition *fuoriserie*, officially sanctioned one-offs, F1 cars, and 488s modified for GT racing and Ferrari Challenge. The only thing missing is a sport-utility vehicle, and that's rumored to be on its way in 2021—at which point Enzo will probably be spinning in his grave faster than a V-12 at redline.

THE PRANCING HORSE EMBLEM

Ferrari's Prancing Horse, *Cavallino Rampante*, is one of the world's most recognizable logos—a rearing black stallion on a field of yellow, with green, white, and red stripes representing the Italian flag at the top and "Ferrari" in a serif typeface at the bottom. Enzo Ferrari claimed that the emblem had been given to him by the mother of Francesco Baracca, Italy's top fighter pilot of World War I. "Ferrari," she told him, "put my son's prancing horse on your cars. It will bring you luck." Ferrari said he appropriated the horse but silhouetted it against a yellow background in honor of Modena, his birthplace. Like most creation myths, this one comes with some inconsistencies. Naysayers point out that, although Ferrari supposedly met with Baracca's parents after a race he won in Ravenna in 1923, the prancing horse shield didn't appear on a car—an Alfa Romeo run by Scuderia Ferrari—until the 24 Hours of Spa in 1932. Why, they reasonably ask, the long delay? Other skeptics suggest that Baracca took the emblem from a German airplane he'd shot down. As it turns out, a prancing horse against a yellow backdrop is also the crest of Stuttgart, Germany, which is why it appears on the Porsche badge. Awkward! Coincidentally, a similar logo used to appear on motorcycles manufactured by Ducati, whose technical director was born in the same town as Baracca. For the record, the Francesco Baracca Museum says Baracca—who was killed after scoring thirty-four victories—personalized the emblem used by an Italian cavalry regiment dating back to 1692. Whatever the truth, the *Cavallino Rampante*—by then in the form of a rectangle rather than a shield—appeared on the first Ferrari in 1947. And it's been on every Ferrari ever since.

GLOSSARY

FERRARI CLASSICHE: Created in 2006, the Classiche department provides restoration services and certificates of authenticity. It also houses an archive full of build sheets covering everything from test data to the configuration of tool rolls.

MILLE MIGLIA: The Mille Miglia, Italian for "One Thousand Miles," was an open-road race staged in Italy from 1927 to 1957. Although the route changed over the years, it generally ran from Brescia to Rome and back. The circuit consisted of public roads that were closed for the day, allowing cars to roar through the center of dozens of towns and cities. From 1948 to 1957, Ferraris won all but two editions of the race.

MONZA: The Autodromo Nazionale Monza is the longtime home of the Italian Grand Prix. Carved out of a forested park near Milan in 1922, it's the second-oldest purpose-built racetrack (after the Indianapolis Motor Speedway) still being used today. Monza has been the site of some of the worst tragedies in Ferrari history. But when a Ferrari wins at Monza, the tifosi unleash a celebration that makes other victory parties seem second rate.

N.V.: Ferrari N.V. is the formal name of the firm—incorporated in Amsterdam—that was spun off from Fiat Chrysler Automobiles in 2016. N.V. are the initials of the Dutch term *naamloze vennootschap*, a publicly traded company. Previously, the firm had been known as Ferrari S.p.A., for Società per Azioni.

SCUDERIA: *Scuderia* means "stable" in Italian. In the motorsports world, it refers to a race team. Numerous teams have incorporated the word into their names—Scuderia Centro Sud in the 1950s, Scuderia Filipinetti in the 1960s and 1970s, Scuderia Corsa to this day. But Scuderia Ferrari is by far the most famous and successful. Ferrari also applied the term to the hot-rodded version of the F430, the 430 Scuderia.

TIFOSI: *Tifosi* is Italian for "fans." Technically, all sports have tifosi. But in racing parlance, the term refers to the rabid—and legion—Scuderia Ferrari devotees who transform grandstands at racetracks all over the world into seas of red. Tifosi need not be Italian, and they shower their passion on anybody in a Ferrari uniform. Thus, long-hated drivers such as Alain Prost and Nigel Mansell were adored during their stints with the Ferrari F1 team—and reviled more than ever after they left.

NINE FOR THE ROAD

1948 FERRARI 166 MM BARCHETTA

FUN FACT

The shape of Touring's simple yet lovely Barchetta inspired several other two-seat roadsters, most notably the AC Cobra that raced against Ferrari in the 1960s.

HISTORICAL TIDBIT

The 166 followed Ferrari's convention of using a model number that represented the individual cylinder capacity in terms of cubic centimeters. Thus, 166 (actually 166.25) multiplied by 12 equals 1995cc, or roughly 2.0 liters.

KEY PERSON

The 166 MM Barchetta displayed at the Turin Auto Show was bought by Tommy Lee, the son of California car and radio tycoon Don Lee, whose employees had included celebrated GM designer Harley Earl and Indy car builder Frank Kurtis.

Enzo Ferrari had already won races in 1947 with his first car, the 125 S, and the 166 that followed it scored historic victories the next year in the Targa Florio and the Mille Miglia. But the first 166s were stocky-looking cars styled by Carrozzeria Allemano, and there was an even more primitive cycle-fender version known as the Spider Corsa. Visitors at the Turin Auto Show were stunned to find a trim roadster of unexpected beauty sitting on the Ferrari stand. Designer (and design historian) Robert Cumberford deemed it "one of the most charismatic sports car shapes ever." Although the car was officially known as the 166 MM, in honor of the Mille Miglia win four months earlier, Giovanni Canestrini, the dean of Italian automotive journalists, dubbed it a *barchetta*—little boat—and the name stuck. The body was styled by Carlo Felice Bianchi Anderloni, the head of Carrozzeria Touring, best known for its patented *superleggera* (superlight) system of coachbuilding. But the Barchetta was more than a pretty face. The car featured a 140-horsepower 2.0-liter version of the 60-degree V-12 that Gioacchino Colombo had designed for the first Ferrari. An all-aluminum single-overhead-cam motor with oversquare dimensions that were revolutionary at the time, the Colombo V-12 proved to be one of the greatest engines in automotive history, and the basic architecture remained in production for the next two decades. Technically, the 166 MM was a race car that could be driven on the street; in fact, it won the Mille Miglia and Le Mans the next year. But Ferrari also offered a detuned road version dubbed the 166 Inter to appeal to a wider audience, and various coachbuilders designed bodies of their own to compete with Touring.

NINE FOR THE ROAD
1959 FERRARI 250 GT SWB

FUN FACT

A twin-nostril version of the SWB, designed by the young Giorgetto Giugiaro for Carrozzeria Bertone and inspired by the shark-nose Formula 1 Ferrari 156, was shown at the Geneva Salon in 1962.

HISTORICAL TIDBIT

After Enzo Ferrari canceled Count Giovanni Volpi's order for a 250 GTO, Volpi hired the GTO's principal designer, Giotto Bizzarrini, to modify a SEFAC hot rod to compete against the GTO. The unusually long, high, and flat rear deck led to the car's not entirely complimentary nickname, the Breadvan.

KEY PERSON

Denise McCluggage, an automotive journalist and one of the preeminent female race drivers of her generation, scored a class win at Sebring in 1961 in a SWB, co-driving with jazz saxophonist Allen Eager.

The 250 GT SWB, for short wheelbase, was the ultimate run-what-you-brung sports racer. Nearly half of the 165 cars built from 1959 to 1962 were designated as Competizione models, with various performance upgrades, and even straight-out-of-the-box street models were quick enough to dominate in club races. The car was an immediate sensation after debuting at the Paris Salon in October 1959. The perfectly proportioned body, drawn by Carrozzeria Pininfarina, evoked the seminal Cisitalia 202 coupe, from the long, flowing hood to the elegant fastback rear deck. The engine was the 3.0-liter version of the venerable Colombo V-12, a design that had been proven in the 250 Testa Rossa. Depending on the state of tune, it made anywhere from 240 to 295 horsepower. The big news was the chassis, which had been shortened 7.9 inches from so-called long-wheelbase models—it now measured 94.5 inches—to sharpen the handling. The SWB was also the first Ferrari to be fitted with tubular shock absorbers and disc brakes as standard equipment. Of course, "standard equipment" is a misnomer when talking about a hand-built car produced in such small numbers. Generally speaking, though, competition versions sported aluminum bodies rather than steel, modified engines, and plexiglass windows. Toward the end of the car's lifespan, in an effort to compete with the new Jaguar E-type, Ferrari developed an even faster model unofficially known as the SEFAC hot rod. (In 1960, Ferrari had been reincorporated as Società Esercizio Fabbriche Automobili e Corse, or SEFAC.) Raced by privateers, SWBs won the Tour de France three times in a row while scoring class wins at Le Mans, Sebring, and the Nürburgring. And they looked—and sounded—fabulous while doing it.

NINE FOR THE ROAD
1962 FERRARI 250 LUSSO

The 250 Lusso wasn't the fastest or most exotic Ferrari. It wasn't made in the largest numbers, nor did it sell for the highest prices. The reason it's remembered so fondly is because it was one of the most beautiful Ferraris ever built. No, check that. It was one of the most beautiful *cars* ever built. Also known as the 250 GT/L, the Lusso—for luxury—was the work of Carrozzeria Pininfarina. It shared its chassis, many of its mechanicals, and its general shape with the 250 SWB, another Pininfarina masterpiece. But because the Lusso was designed to be a civilized grand touring machine rather than a race car, it benefited from several cosmetic refinements. At the front end, the fenders were extended, and an unusual three-piece bumper was added. The humps over the rear wheels were eliminated, flattening the rear haunches. Windows were installed between delicate B and C pillars, creating an airy greenhouse. And yet the car was almost devoid of ornamentation. As a result, the Lusso exuded the rigorous grace of a ballerina rather than the brawny charisma of the SWB. Because the car was meant strictly for cruising, the engine—the ubiquitous Colombo V-12 downgraded to produce a relatively sedate 240 horsepower—was moved forward to provide more room for the car's two occupants. The spacious interior featured leather bucket seats, sound-deadening material, a wood-rimmed Nardi steering wheel, and a unique instrument panel with the speedometer and tachometer in the middle of the dashboard, angled toward the driver. The Lusso was the last of the 250-series road cars that dated back more than a decade, so it seems only fitting that it should be the loveliest of the breed.

The 275 GTB was the first road-going Ferrari to be fitted with a fully independent suspension. It was the last to carry the Colombo-derived V-12 that had defined Ferrari since the marque was launched. The bold, sensuous berlinetta was therefore both a beginning and an end. The Pininfarina-styled coupe debuted at the Paris Auto Salon in 1964 with classic long-hood, short-deck proportions that evoked the 250 GTO. (A less dramatic, more refined spider, dubbed the 275 GTS, was shown at the same time.) The engine had been bored out to 3.3 liters to generate more low-end torque and produce 280 horsepower. The other technical highlight was the incorporation of a transaxle—a unit integrating the gearbox and differential—located at the rear to improve the weight distribution. But options abounded. Three carbs or six. Campagnolo alloy rims or Borrani wire wheels. Steel body or aluminum. Although the car had been designed for the road, Ferrari expected some owners to race it, and it was duly homologated for GT competition. In fact, a 275 painted in the bright yellow livery of the Belgian Ecurie Francorchamps team finished first in class—and an amazing third overall—at Le Mans in 1965. At the end of 1965, Ferrari introduced a long-nose version developed to reduce front-end lift at high speed. (The fact that it was also more attractive was an unintended consequence.) But the ultimate evolution of the 275 was the GTB/4, which arrived in 1966. The big news here was the engine, a free-breathing dry-sump four-cam screamer that ratcheted up performance to 300 horsepower at 8,000 rpm. A flame burns brightest just before it dies.

NINE FOR THE ROAD
1969 DINO 246 GT

FUN FACT

Fiat took control of Lancia as well in 1969, so the Dino 246's drivetrain was also used in the jaw-dropping Lancia Stratos, which dominated international rally racing during the mid-1970s.

HISTORICAL TIDBIT

Fiat, which assumed responsibility for Ferrari's road car business in 1969, produced its own front-engine versions of the Dino—a spider bodied by Pininfarina and a 2+2 by Bertone.

KEY PERSON

Although Enzo credited his son Dino with designing the V-6 engine named in his honor, it was primarily the work of master engineer Vittorio Jano, who'd joined Ferrari after the Lancia Formula 1 team folded in 1955.

Ferrari's own brochure for the Dino, the company's first midengine road car, damned it with faint praise: "Tiny, brilliant, safe . . . almost a Ferrari." Ouch! In fact, the Dino 246 GT is among the marque's best-loved cars—a sexy yet athletic two-seater offering the highest cornering limits of the day at a relatively affordable price. But nowhere on the curvaceous body was the car identified as a Ferrari. Dino was the name Ferrari applied to V-6 engines in honor of his son, Alfredo, a.k.a. "Alfredino" or "Dino," who'd been working on the first version of the motor when he died in 1956. By the mid-1960s, a compact V-6 seemed like a perfect fit for a small new sports car that Enzo envisioned to compete with the Porsche 911. The Dino 206 GT—for 2.0 liters and six cylinders—went into production in 1968 with the engine mounted transversely atop the gearbox and differential between the cockpit and the rear axle. Initially, the car had a lovely aluminum body fashioned by Pininfarina and built by Scaglietti. But the next year, the definitive longer-wheelbase, steel-bodied iteration of the car appeared. Known as the Dino 246 GT, it featured an iron-block 2.4-liter engine rated at 195 horsepower. Top speed was "only" 150 miles per hour. But with a fully independent suspension and a low polar moment of inertia, the car was more agile than anything else on the road. "It makes all front-engine or rear-engine cars obsolete," Denis Jenkinson declared in *Motor Sport*. In 1972, a targa-top model designated the GTS was introduced. Although the Dino turned out to be too labor intensive and expensive to compete with the 911, it became Ferrari's bestselling model to date.

There aren't too many duds in the Ferrari portfolio, but the Dino 308 GT4 merits a spot in the hall of shame. Envisioned as a (relatively) high-volume car, it arrived in 1973 as a 2+2 with a new V-8 engine—the first in a Ferrari road car. But the angular styling by Bertone was unloved, and its performance was underwhelming, so the wedge-shaped Ferrari 308 GTB was rushed into production in 1975. The midengine berlinetta featured the GT4's 3.0-liter V-8 engine—rated at 255 horsepower, or 240 in US spec—in a smaller and much more striking two-seat package. The handsome fiberglass body, styled by Leonardo Fioravanti at Pininfarina, evoked the Dino 246 and established the design language for the next generation of street cars. Like all automakers during the bad old days of the 1970s, Ferrari struggled to adapt to new safety and emissions standards, so the 308 was the subject of numerous upgrades over the years. The fiberglass bodies were junked in favor of steel by 1977, which also saw the debut of a targa-top spider, the GTS. Next, the quartet of Webers was replaced by Bosch K-Jetronic fuel injection, which reduced horsepower along with emissions. (A small "i" was appended to the model name.) To claw back some lost performance, four-valve versions known as Quattrovalvoles were introduced in 1982. Three years later came the final evolution of the car, the 3.2-liter 328 GTB/GTS, which benefited from more power, an updated cockpit, and freshened bodywork incorporating harmonious color-keyed bumpers. Thanks in part to a starring role on the television show *Magnum, P.I.*, nearly twenty thousand 308s and 328s were built by the time production ceased in 1989, shattering the company's previous sales records.

NINE FOR THE ROAD
1984 FERRARI TESTAROSSA

FUN FACT

There were actually two Testarossas on the *Miami Vice* set, both painted white because director Michael Mann thought they'd look better in night shots. But stunts were done with a fiberglass fake built on a De Tomaso Pantera chassis.

HISTORICAL TIDBIT

Testarossa sales were banned in California in 1991 because the car wasn't equipped with on-board diagnostics. This flaw was addressed the following year with the 512 TR, which debuted at the Los Angeles Auto Show.

KEY PERSON

Although consumers clamored for a drop-top version, only one officially sanctioned spider was built—for Gianni Agnelli. It's good to be the king, or, in this case, the head of Fiat and one of Italy's richest men.

The Testarossa was the perfect avatar of the Excessive Eighties, and not just because it was a hero car on *Miami Vice*. Everything about it was over the top—the booty-call rear end, the "egg-slicer" side strakes, the outrigger side-view mirror, and, of course, the magnificent flat-twelve engine. Originally developed for Formula 1, the motor had already done street-car duty in various iterations of the Berlinetta Boxer. But none of these models met American safety and emission standards, so the Testarossa was developed as a world car with a special focus on the US market. The body was styled and built by Pininfarina (rather than Scaglietti) after spending countless hours in the company's full-scale wind tunnel. The mid-mounted radiators solved cooling issues endemic to the BBs while providing more room for luggage, and the car's capacious dimensions—it was nearly seven inches wider than the contemporary Corvette—gave occupants plenty of room to stretch out in the luxurious cockpit. The 4.9-liter engine benefitted from a new four-valve head decorated with scarlet valve covers, hence the name Redhead. (It's not clear why the car was named Testarossa rather than Testa Rossa, the name used in the 1950s.) Besides making glorious noises, the flat-twelve produced 390 horsepower and reached a top speed of 180 miles per hour. The styling was (and remains) polarizing; Sergio Pininfarina himself called it "an exaggeration in flamboyance." But it was a smash hit with customers, who made it one of the most popular Ferraris ever built. The Testarossa was overhauled in 1992 and dubbed the 512 TR. Three years later, the F512 M marked the end of the road for Ferrari's flat-twelve street cars.

FUN FACT

With 109 horsepower per liter, the V-8 in the F355 set a record for normally aspirated street cars. No wonder Frère called it "probably the best sports car engine ever made."

HISTORICAL TIDBIT

On the other hand, building the motors with cogged timing belts means that an engine-out service—priced at between $6,000 and $10,000—has to be performed every three to five years. *Ka-ching!*

KEY PERSON

Peter Sachs won the inaugural US F355 Challenge championship in 1995, but Derek Hill—Phil's son—claimed the world title in the Ferrari Challenge International Finale at Mugello.

The 348 was supposed to be Ferrari's entrée into a brave new world of monocoque construction and modern electronics. So much for the best-laid plans. Introduced in 1989, the 348 was savaged by critics who called it unreliable, rough riding, and hard to handle. "This was clearly the worst product Ferrari had developed in some time," Luca di Montezemolo said, speaking as an unimpressed 348 owner. One of his first acts after becoming president of Ferrari in 1991 was greenlighting a replacement designed to be robust enough to use as a daily driver. The magnificent F355 was the result, and it would be remembered as one of the finest, as well as one of the most important, cars in company history. The F355 was similar to the 348 in many respects, but better in every way. The 3.5-liter twin-cam V-8, which sat in a tubular subframe bolted to a central monocoque, made 380 horsepower at 8,250 shrieking rpm thanks to a flat crank, five valves per cylinder, and titanium connecting rods. The impeccably proportioned body-work by Pininfarina (natch) prompts many modern observers to call the F355 the last classically styled Ferrari. Even more dramatic upgrades were found inside the car. Ferrari glammed up the interior with Connolly leather and bespoke switchgear; previously, components had been raided from the Fiat parts bin. Also, the example of the user-friendly Acura NSX, which arrived in 1990, compelled Ferrari to overhaul its cockpit ergonomics. The F355 debuted in 1994 in berlinetta and GTS (removable targa top) form, and a Spider appeared the next year. *Road & Track*'s Paul Frère called the F355 "the purest purebred yet from Ferrari's scuderia." High praise indeed from the guy who won Le Mans in 1960—in a Ferrari 250 Testa Rossa.

NINE FOR THE ROAD
2012 FERRARI F12BERLINETTA

The design brief was deceptively simple: Create the fastest, most powerful series-production Ferrari ever. But there was a catch: It also had to be reliable and comfortable enough to use as a daily driver. The result was the F12berlinetta, and if the car had a fault, it was the perverse model name. In some respects, the F12 was a throwback to the front-engine, rear-transaxle configuration pioneered in the 275 GTB in 1964. The chassis was old-school aluminum rather than cutting-edge carbon fiber. But the car was significantly shorter, lower, narrower, lighter, and stiffer than the 599 GTB Fiorano that it replaced. Better still, it was powered by the latest in Ferrari's long line of normally aspirated V-12s. The plus-sized 6.3-liter motor shared its general architecture with the powerplant in the FF—Ferrari's somewhat ungainly four-seater—but it was tweaked with a new block, freer-flowing exhaust and intake, more aggressive cams, and a mind-bending compression ratio of 13.5:1. The upgrades added up to 731 horsepower, an engine capable to spinning to 8,700 rpm and a top speed of 211 miles per hour. Keeping the car on the road was a serious concern, so it was full of trick aerodynamic aids, and the body—designed in-house at the Ferrari Styling Centre rather than Pininfarina—was so deeply sculpted to direct air in the right directions that it looks pugnaciously functional rather than drop-dead gorgeous. But the cockpit was sufficiently luxe and comfy to satisfy the most demanding sybarites, while the seven-speed dual-clutch gearbox, five-link rear suspension, adjustable dampers, and lockable electronic diff made the car a sweetheart anywhere from the autostrada to the Autodromo Nazionale Monza.

WORST FERRARIS EVER

Dedicated fanboys and PR functionaries who've drunk too much corporate Kool-Aid claim that Ferrari never made a genuinely bad car. Really? What about the stumpy Mondial, a fickle 2+2 that could be adored only by a blind man (Al Pacino in *Scent of a Woman*)? Or the 348, which even Ferrari president Luca di Montezemolo called a piece of junk? (It probably sounded more poetic in Italian.) Let's face it: There are a handful of Ferraris that deserve jeers rather than cheers. Just about everybody's favorite whipping boy is the 400/412, a bloated businessman-mobile often fitted with a three-speed BorgWarner slush box. On the racing front, it's hard to top the F14 T Formula 1 car, which didn't win a single race despite being driven by a pair of World Champions, Fernando Alonso and Kimi Räikkönen. Plus, it looked like a mutant anteater. There's also a long history of bizarre one-offs and show cars. The 250 GT SWB-based Breadvan was an eyesore, though it was reasonably competitive. But the only positive comment that can be made about the Bertone Rainbow, an origami-shaped special drawn by Lamborghini Countach stylist Marcello Gandini around a Dino 308 GT4 platform, was that it never went into production. A special place in automotive purgatory is reserved for kit cars featuring fiberglass bodies that vaguely resemble Ferraris mounted on decidedly non-Ferrari chassis. Virtually every model from the 250 GTO to the Enzo is available in faux-Ferrari form. Some are nicely done. The vast majority are not. Donor cars include Celicas, Datsun Zs, Mercury Cougars, and several generations of Corvettes. But the most infamous replicars start with Toyota MR2s and Pontiac Fieros, because their midengine layout works best for impersonating modern Ferraris.

GLOSSARY

BERLINETTA: Although berlinetta translates as "little sedan," it refers in Ferrari vocabulary to a two-door coupe with a sporty character. Larger two-door coupes built as 2+2s—that is, with back seats roomy enough to fit two (small) people—don't carry the berlinetta designation.

CHASSIS: The chassis is the platform that forms the base of a car. For decades, most cars were built around ladder-style frames to which the body was mounted—the so-called "body-on-frame" design. Modern chassis generally feature unibody construction in which the frame, floor, and body are a single piece—hence the shortened form of the term "unit-body"—which enhances structural rigidity while saving weight.

MIDENGINE CHASSIS: Traditional cars, including Ferrari race cars until the 1960s and street cars until the 1970s, feature an engine in front of the cockpit. Rear-engine chassis place the motor behind the rear axle line; the Porsche 911 is a rare example. Most modern race cars and high-end sports cars incorporate a midengine chassis, with the powerplant nestled between the rear axle line and the cockpit. While it compromises interior space, this midengine architecture makes the car more agile.

OVERHEAD CAMSHAFT: While less sophisticated overhead-valve engines use pushrods to open the valves, overhead camshafts actuate the valves directly. A single-overhead-cam engine has one camshaft for each bank of cylinders. A dual-overhead-cam or twin-cam motor has two. This allows the cylinder head design to be optimized, which improves engine breathing, raises the redline, and, ultimately, increases power.

SPIDER: Spider is the term used to describe roadsters and convertibles. In many cases, it isn't part of the official Ferrari name. Although there is no "Y" in the Italian alphabet, it's often misspelled as "spyder." But even Ferrari—which now uses "spider" exclusively—called the famous Scaglietti drop-top a "250 California Spyder" in its own sales literature.

SUPERLEGGERA: *Superleggera*, Italian for "superlight," was a coach-building system patented by Carrozzeria Touring. It entailed fabricating a frame of small-diameter, lightweight, steel tubes to which thin, hand-formed aluminum body panels were welded. After the 166 MM Barchetta, the technique was later used in several Aston Martins and Lamborghinis.

EXCLUSIVE, EXOTIC, EXHILARATING

EXCLUSIVE, EXOTIC, EXHILARATING
1968 FERRARI 365 GTB/4 DAYTONA

FUN FACT

In the 1980s, Daytona spiders were worth so much more than berlinettas that about one hundred coupes were cut down into convertibles.

HISTORICAL TIDBIT

In 1979, John Morton and Tony Adamowicz finished a remarkable second overall in the 24 Hours of Daytona—ahead of nearly a dozen Porsche 935s—in a 1973 365 GTB/4 Competizione originally owned by Luigi Chinetti.

KEY PERSON

In 1971, Dan Gurney and Brock Yates won the Cannonball Baker Sea-to-Shining-Sea Memorial Trophy Dash—an unsanctioned race across the continental United States better known as the Cannonball Run—in a stock Daytona. "At no time did I exceed 175 miles per hour," a smiling Gurney said afterward.

Depending on your perspective, Enzo Ferrari was either a principled traditionalist or a stubborn old coot. His decision to delay switching to rear-engine race cars cost him countless Grand Prix and sports car victories. On the other hand, his refusal to ape the midengine Lamborghini Miura produced one of the most memorable and highly admired Ferraris—the 365 GTB/4, known universally (except by Ferrari) as the Daytona. With a long, bold hood shrouding a Colombo-derived V-12, the Daytona was the last production Ferrari to be assembled by hand and Enzo's final gesture before turning over the road car side of the business to Fiat. As soon as it appeared, *Road & Track* anointed it "the best sports car in the world. Or the best GT. Take your choice; it's both." The car was the successor to the 275 GTB/4, with which it shared its basic platform and most of its mechanical components. But the body, styled by Leonardo Fioravanti at Pininfarina, represented a major departure—more angular and modern than previous Ferraris, with retractable headlights defining the leading edge of the shark-like prow. A sweet limited-production spider known as the 365 GTS/4 broadened the Daytona's appeal. Styling aside, the car's most impressive feature was a 4.4-liter, 352-horsepower, four-cam engine that produced endless midrange torque and wound out to 174 miles per hour. A race version of the Daytona proved to be worthy of its nickname, which referred to Ferrari's line-abreast 1-2-3 finish at the 24-hour enduro in 1967. Fifteen competition cars were sold through the Assistenza Clienti department in Modena and raced by privateers who scored several class wins at Le Mans and won the Tour de France outright in 1972.

EXCLUSIVE, EXOTIC, EXHILARATING
1973 FERRARI 365 GT4 BB

Enzo Ferrari insisted on doing things his way. When it was time to replace the front-engine Daytona, he realized he had to build a midengine sports car to compete with the Lamborghini Miura, the Maserati Bora, and the coming Countach. And he did. But instead of designing it around a V-8 or even a V-12, he created the world's first flat-twelve production car. Officially designated the 365 GT4 BB, it was popularly known as the Berlinetta Boxer due to the horizontally opposed cylinders. At 4.4 liters, it was the same size as the engine in the Daytona. And, in fact, it was more like a flattened version of the Daytona motor than a road-going take on the flat-twelves being raced at the time in Ferrari's Formula 1 and sports-prototype cars. The factory rated the engine at 360 horsepower and quoted a top speed of 188 miles per hour, but these claims were laughably optimistic. The Boxer was the first Ferrari with timing belts rather than chains, which reduced cost and noise but meant more maintenance down the road. A 5.0-liter dry-sump version dubbed the 512 BB debuted in 1976 and a fuel-injected model, the 512 BBi, arrived in 1981, but these provided better tractability without improving performance. The body—the last to be hand-formed by Scaglietti—was a delicate, handsome Pininfarina design often featuring a striking two-tone paint scheme delineating the upper and lower halves of the car. To improve the packaging, the engine sat on top of the transaxle, and a space-saver spare tire was provided. Even so, room for occupants and storage was notoriously limited, and the Boxer never developed a fan club as fanatical as the Daytona's.

EXCLUSIVE, EXOTIC, EXHILARATING
1984 FERRARI 288 GTO

The 288 GTO was a bullet that missed its original target. But Ferrari then shifted its aim, and the modern GTO scored a bull's-eye, creating a market for limited-edition halo models that are now known as supercars. In theory, at least, the car was conceptualized as an entry in the World Rally Championship's Group B category. The name, which traded on the reputation of the iconic 250 Gran Turismo Omologato of the 1960s, was chosen because two hundred cars had to be built to homologate the 288 GTO for WRC competition. To save time and money, Ferrari based it on the 308 GTB. Pininfarina's Leonardo Fioravanti was responsible for styling the 288 GTO, which looked like the 308's more muscular older brother and is often considered the most attractive Ferrari of the 1980s. Engineer Harvey Postlethwaite, who'd been brought in to revive Ferrari's Formula 1 program, reoriented the engine by 90 degrees to face north–south. This allowed the transaxle to be mounted longitudinally, making it easier to change gear ratios. It also provided space for a pair of small turbos along with their plumbing. Yes, this was Ferrari's first turbocharged street car engine, and it really rocked. Based on the turbo V-8 in the blazing-fast Lancia LC2 Group C car (which, in turn, had been derived from the engine in the normally aspirated Ferrari 308), the 2.8-liter motor developed 400 horsepower, giving the 288 GTO a top speed of 190 miles per hour. Kevlar-reinforced composite body panels—the first ever used in a road car—minimized weight and enhanced performance. Group B fizzled in 1986, but who cared? By that time, Ferrari had already presold all 272 cars, rendering the 288 GTO an instant collectible.

EXCLUSIVE, EXOTIC, EXHILARATING
1987 FERRARI F40

FUN FACT

Like the Model T Ford, the F40 came in only one color: red. The single piece of trim in the cockpit was gray cloth covering the instrument panel. The only creature comfort was air conditioning. Options? There were no options.

HISTORICAL TIDBIT

About twenty race models designated as the F40 LM (the first two, which raced at Le Mans) and the F40 Competizione (all the rest) were developed with Michelotto Automobili. Although the cars won some minor races, they were never front-line contenders.

KEY PERSON

Nicola Materazzi, a protean engineer who had played a principal role in the creation of the Lancia Stratos and 288 GTO, is often called the father of the F40.

The F40 is often considered Ferrari's first truly modern supercar, and why not? Carbon-kevlar body panels minimized weight, and twin turbochargers boosted the speed up to 200 miles per hour. The bodywork featured more NACA ducts than a supersonic fighter jet, and the integrated rear wing was large enough to obliterate the sight of any neighboring Lamborghini Countaches. Despite the hype, the F40 actually represented the end of an era. It was very much the successor to the 288 GTO, and it was the last road car approved by Enzo Ferrari. When it debuted in 1987, marking the fortieth anniversary of his company, he told the crowd in Maranello, "Little more than a year ago, I expressed my wish to the engineers. Build a car to be the best in the world. And now the car is here." Not everybody agreed. Groundbreaking designer Gordon Murray dissed the F40 because he considered its chassis antiquated, and everybody agreed that the contemporary Porsche 959 was far more sophisticated technologically. But savage simplicity was precisely what the F40 offered. It was the most impractical and uncompromising Ferrari road car since the 1950s. There were no power brakes, no power steering, no upholstery, no door handles, no glovebox—no features that didn't make the car go faster. With a 2.9-liter engine rated at 478 horsepower and a curb weight around three thousand pounds, the F40 was the fastest car *Car and Driver* had ever tested. (It also posted the highest lateral-acceleration number on a skid pad: 1.01 g.) Initially priced at about $400,000, some models brought more like $900,000. And, even though 1,311 were built, they still sell for more than $1 million today.

EXCLUSIVE, EXOTIC, EXHILARATING
1995 FERRARI F50

Piero Ferrari was the inspiration for the F50, the supercar celebrating the marque's fiftieth anniversary. According to company lore, he used an F40 as a daily driver, and he wanted its replacement to be every bit as advanced and uncompromising. It was only natural that the new car should incorporate elements from Formula 1. The semi-automatic paddle-shift transmission didn't transfer over because Ferrari didn't think the electronics were robust enough for road use. But the F50's 4.7-liter, 520-horsepower V-12 engine was an iron-block take on the powerplant found in the race-winning F1-89, and the carbon fiber tub and suspension were straight out of the formula-car playbook. The styling was clearly in the vein of the F40, with prominent air ducts and a gigantic, fully integrated rear wing. Although Pininfarina wanted to offer a coupe and a roadster, Ferrari split the difference with a single dual-purpose model with a removable targa top. The F50 debuted during the uncertain years after the recession of the 1990s, so production was limited to 349—one fewer than perceived demand—and, instead of selling cars, they were leased at what many people thought were rapacious rates. (In the United States, the down payment was $240,000, with twenty-four monthly payments of $5,600 and a buyout price of $150,000.) Reviews of the styling were mixed. Ditto for the performance. Despite the F1 technology, the V-12 lacked the propulsive punch of the twin-turbo V-8 in the F40. Even worse, the car wasn't as fast or sophisticated as the McLaren F1. Plans to build a race version of the F50 were stillborn. But never mind. The street car—a scarlet hot rod in supercar clothing—was more visceral than anything on the road.

EXCLUSIVE, EXOTIC, EXHILARATING
2002 FERRARI ENZO FERRARI

Critics damned the F50 with faint praise. Superlative, yes, but not awe inspiring. Ferrari was determined not to make that mistake again. "I wanted to go a little bit too far in every element to build a super-extreme car," Ferrari president Luca di Montezemolo said. Mission accomplished with the Enzo. (Technically, the car was named the Ferrari Enzo Ferrari, but nobody outside Maranello used this pointlessly redundant construction.) The angular, in-your-face body looked like the product not of a design studio but a wind tunnel. Love it or hate it, the Enzo was an aerodynamic marvel. In addition to a moveable rear wing, flaps underneath the body directed air to the diffuser tunnels. Downforce maxed out at 1,709 pounds at 186 miles per hour before dwindling progressively until the car reached a breathtaking top speed of 217.5 miles per hour. Power—660 ponies to be exact—came from a 6.0-liter V-12 hot-rodded with Nikasil-lined cylinders and titanium connecting rods. The engine, the first of a new family of normally aspirated V-12s, was mated to a Formula 1–derived, six-speed, semi-automatic transmission that, in Race mode, shifted as subtly as a sledgehammer. The tub was carbon fiber. So were most of the body panels, including the scissor doors, as well as the flat-top steering wheel. The car also featured a formula car–style pushrod suspension and the first production-car use of carbon-ceramic brakes. There was no audio system, no navigation system, no cruise control. Air conditioning was optional. Not enough creature comforts for you? Tough noogies. All 349 cars in the initial production run were sold before they were built. Ferrari then cranked out fifty-one more, and those were snapped up just as quickly.

EXCLUSIVE, EXOTIC, EXHILARATING
2014 FERRARI LAFERRARI

FUN FACT

Unlike the systems in the other hybrid supercars on the market, the Porsche 918 Spyder and McLaren P1, the Ferrari's electric motor was calibrated only to improve performance, not fuel economy.

HISTORICAL TIDBIT

In 2017, the 210th and last LaFerrari Aperta sold at auction for $10 million, the highest price ever paid for a car built in the twenty-first century. The proceeds were donated to Save the Children.

KEY PERSON

Technical director Roberto Fedeli was responsible for engineering the LaFerrari. After it launched, he decamped to BMW before returning to the FCA fold in 2016 as technical director of Maserati and Alfa Romeo.

When each new Ferrari supercar is unveiled, it's inevitably hailed as the greatest Ferrari ever. Which makes sense. If nothing else, the new car should be at least incrementally better than the one it replaced. But the LaFerrari—Ferrari's first road-going hybrid—made a quantum leap over the Enzo. The car's signature feature was a 161-horsepower electric motor running through a Formula 1–inspired Kinetic Energy Recovery System (KERS). Coupled with a 6.3-liter normally aspirated V-12 rated at 788 horsepower, the hybrid power unit made a grand total of 949 horsepower, with 663 pounds-feet of torque available almost from rest. Top speed was 218-plus miles per hour—roughly the same as the Enzo—but the LaFerrari was a whopping five seconds a lap faster at Fiorano. The car was built around a Rory Byrne–designed carbon fiber tub into which the seat was molded to lower the center of gravity by 1.3 inches. Because the seat was fixed, the steering wheel and pedal box moved. (Owners were also fitted for custom seat padding.) Onboard computers controlled everything from the electronic diff to the magnetic dampers to the active aerodynamics, but all the systems were so seamlessly integrated that they were transparent to the driver. The car was the first Ferrari to be styled entirely in-house, and the first since the Bertone-designed Dino 308 GT4 in 1973 with no input from Pininfarina. All 499 cars in the initial build were presold for somewhere around $1.5 million apiece. (The 500th was auctioned off for charity.) Ferrari then made 209 LaFerrari Apertas, also presold, with removable carbon fiber tops. Rumor has it that LaFerraris were offered only to favored customers who'd recently bought two Ferraris through dealerships and who had owned six in the previous ten years.

ONE-OF-A-KIND FERRARIS

Unlike mainstream car companies, Ferrari has never gone gaga over show cars. Maybe it's because the production cars were already so exotic and built in such small numbers that there was no need to impress anybody. Still, Ferrari has had its share of wild, futuristic, and just plain bizarre one-off models that never made it into showrooms. One of the most memorable was the 410 Superamerica outfitted by Ghia in 1956 with huge tail fins, chrome galore, and a wraparound windshield that would have been at home on an over-the-top American car of the era. Perhaps not coincidentally, this was the last Ferrari designed by Ghia. Kudos to Bertone for a lithe and lovely 250 GT SWB with twin nostrils inspired by Ferrari's contemporary Formula 1 car. Demerits to Vignale for transforming a 330 GT 2+2 into a station wagon. No design house has been more closely associated with Ferrari than Pininfarina, which produced a remarkable quartet of vastly different, though similarly named, show cars between 1968 and 1970: the aerodynamic but ungainly P5; the low, slender P6, which developed the design language for the Berlinetta Boxer; the wedge-shaped 512 S Berlinetta Speciale; and the 512 S-based Modulo, which looked like it had driven off the set of a sci-fi movie. There's also a long tradition of one-off Ferraris commissioned by wealthy clients, such as the 375 MM spider that film director Roberto Rossellini had Scaglietti modify into a coupe for his wife, Ingrid Bergman, in 1955. In recent years, Ferrari has created one-offs of its own through the Ferrari Styling Centre. An early project, the P540 Superfast Aperta, was a modern riff on the fantastic Fantuzzi-bodied 330 LMB spider of 1964. Another winner was a 458 Italia-based coupe designated as the SP12 EC—that's "EC" as in Eric Clapton, the car's owner.

GLOSSARY

CARBON FIBER: Carbon fiber is the generic term used to describe composite materials typically formed by combining a weave of carbon filaments with a resin reinforcement. Components made of carbon fiber are lighter and stronger than even the most expensive and exotic metals. Originally developed for the aerospace industry, they were introduced to the racing world in the 1980s and are now found on high-end sports cars.

DOWNFORCE: Downforce is the aerodynamic thrust pressing a car to the ground, increasing cornering power. Downforce can be produced by a splitter on the nose, a spoiler or wing at the rear, and diffuser tunnels underneath the car. The drawback of downforce is increased drag. Moveable aerodynamic aids allow higher levels of downforce for cornering while reducing drag for straight-line speed.

ENGINE ARCHITECTURE: Internal combustion engines are characterized by the number and configuration of their cylinders, typically in the shape of a "V" or an "I." The engine most commonly associated with Ferrari is the Colombo V-12, which was built around two banks of six cylinders arranged in a 60-degree V. The flat-twelve engine was technically a 180-degree V. The only non-V engines developed by Ferrari were the Lampredi I-4 (also known as an inline-four or straight-four) and I-6 (ditto).

GEARBOX: The gearbox is a device that transmits power from the engine to the driven wheels. Early cars were equipped with manual gearboxes that required a clutch to shift gears. Automatic transmissions, popularized in the 1950s, substitute a torque convertor for the clutch. A semi-automatic, pioneered by Ferrari in the paddle-shift 640 Formula 1 car in 1989, uses electronics to pneumatically actuate the clutch. Many street cars are now equipped with dual-clutch transmissions, which function like more civilized semi-automatic gearboxes.

TURBOCHARGING: Turbocharging is a form of forced induction that captures exhaust gases and uses a turbine to pump them back into the engine to boost power. This added efficiency allows automakers to use smaller engines that consume less fuel and produce fewer emissions. Supercharging—an alternate form of forced induction—was used on the first Colombo V-12.

DESIGNED FOR WINNING

DESIGNED FOR WINNING
1952 FERRARI 500

Irony alert! The first Grand Prix Ferrari to win a World Championship had four cylinders rather than twelve. And it was designed not by Gioacchino Colombo, whose V-12 is an essential element of the Ferrari legend, but by the underappreciated Aurelio Lampredi. In fact, it was Lampredi rather than Colombo who created the unsupercharged 4.5-liter V-12 that finally subdued the Alfa Romeo 159 in 1951. For his next engine, Lampredi opted for four cylinders instead of twelve. Why? Because, nearly one hundred pounds lighter, its power-to-weight ratio was superior. Because, with 65 percent fewer reciprocating parts, it was more reliable. Because bigger cylinders and a longer stroke translated into more midrange torque. The 2.0-liter inline-four was designed for Formula 2 competition. But when the beaten Alfettas withdrew from racing, Formula 1 withered on the vine, and Formula 2 was unexpectedly elevated to the premier level of Grand Prix competition. The Ferrari 500, as Lampredi's tidy masterpiece was dubbed, was the class of the field. Although the ladder-style frame was standard-issue chassis design, twin radius rods and a De Dion rear end improved the car's road-holding qualities. Better still, Scuderia Ferrari had the incomparable Alberto Ascari as its team leader. He won every GP he contested in 1952, and his teammate, Piero Taruffi, won the only one that Ascari missed (Ascari was off racing a Ferrari with a Lampredi V-12 at the Indy 500). The next year, the Ferrari 500 again won every championship race but one—a fierce slipstreaming battle at Monza, where Juan Manuel Fangio prevailed in a Maserati—and Ascari earned his second consecutive championship. Lampredi's four-bangers turned out to be a technical dead end. But for two years, they were unbeatable.

DESIGNED FOR WINNING
1958 FERRARI 250 TESTA ROSSA

The original idea was to sell 250 Testa Rossas to rich amateurs as a moneymaker. Instead, the cars became factory racers that won Le Mans four times. That, as they say, is racing. The Testa Rossa name—for "red head," referring to the color of the camshaft covers—was first used in 1956 on the 2.0-liter, four-cylinder 500 TR, which was raced solely by gentleman drivers. By 1957, Ferrari thought it might be a good idea to upgrade the car with a 3.0-liter V-12 engine. By this time, the company was racing exotic four-cam versions of the V-12. But in the interest of durability, chief engineer Carlo Chiti reverted to the old, reliable Colombo design, though he modified it with modern touches—such as double-coil valve springs—that allowed the motor to remain in production for six more years. When the FIA announced that, starting in 1958, the top class of sports car racing would be limited to 3.0 liters, Ferrari decided to build TRs both as factory entries and for customers. Sergio Scaglietti fashioned and built striking pontoon-fendered bodywork inspired by Formula 1 designs. Works Ferraris finished 1–2 when they debuted in Buenos Aires in 1958 and waltzed to the championship. The next year, the aerodynamically disastrous pontoon fenders were replaced by more conventional bodywork designed by Pinin Farina and built by Carrozzeria Fantuzzi. Two more championships followed in 1960 and 1961 with the TRI 60 and TRI 61. (The "I" referred to the improved independent rear suspension.) A 4.0-liter mongrel dubbed the 330 TRI even managed to win Le Mans in 1962, spanking the newer, more sophisticated rear-engine Ferrari Dinos. It was the last front-engine car to prevail at Le Mans.

DESIGNED FOR WINNING
1961 FERRARI DINO 156

The Dino 156 wasn't Ferrari's best Formula 1 car, but it was among its winningest. It was also one of the most attractive, with a highly stylized twin-nostril shark nose and a shapely but classically restrained body hammered out by Fantuzzi metalworkers. The Dino 156 was the first rear-engine Ferrari to go into production—belatedly, because Enzo had long resisted the notion of moving motors from their traditional spot in front of the cockpit. Despite its modern configuration, the chassis was beefy and crude in comparison with its rivals. But the Dino 156 was in a league of its own on the engine front. A new 1.5-liter formula went into effect in 1961. While the British constructors were caught napping, Ferrari built on the success of its proven Formula 2 motor. A 65-degree V-6 designed by Vittorio Jano was already winning races on the junior circuit, and technical director Carlo Chiti developed a more compact, better-balanced 120-degree version for F1. The 190-horsepower engine was so much more potent than the competition that the car won in its debut, at the non-championship Syracuse Grand Prix, even though its driver, Giancarlo Baghetti, was making his first F1 start. Ferraris finished 1-2-5 at Zandvoort, 1-2-3-4 at Spa, and 1-2-3 at Aintree while winning five of seven World Championship events. Only the genius of Stirling Moss at Monaco and the Nürburgring prevented a clean sweep. American Phil Hill won the driver's title, with teammate Wolfgang von Trips a single point behind him despite having been killed in the Italian Grand Prix. With an upgraded chassis and conventional bodywork, the Dino 156 continued to win occasionally through 1964. But the shark-nose version is the model that looms largest in Ferrari lore.

DESIGNED FOR WINNING
1962 FERRARI 250 GTO

FUN FACT

After its race career ended, one GTO was donated to a high school shop class in Victoria, Texas. It later spent more than a decade on an open trailer in a grass field before being restored and displayed at Pebble Beach.

HISTORICAL TIDBIT

Another GTO was seized from the estate of a murdered drug dealer who reportedly had bought the car with cash carried in a backpack. It was sold by the FBI in 1987 for $1.6 million.

KEY PERSON

French gold medal skier Henri Oreiller was the only driver killed in a GTO, impaled on the gearshift lever after crashing at the Montlhéry Autodrome in 1962. In 2014, the car brought $38.1 million at auction—the highest price ever paid for a car in a public sale.

The 250 GTO marked the apogee of the first epoch of Ferrari history, and it distilled the finest attributes of the brand into a single model—performance, passion, panache, power, grace, and charisma. The car has been described as a refined 250 SWB or a 250 Testa Rossa with a roof, and it's true that it was an extension of what came before rather than a break with the past. But the GTO so thoroughly incorporated the lessons learned from its predecessors that it approached perfection, winning twenty of twenty-eight races and claiming three consecutive championships. GTO stood for *gran turismo omologato*, which was, in a sense, exactly what it wasn't. According to FIA regulations, one hundred cars had to be built to homologate a model for GT competition. In fact, only thirty-nine GTOs were produced, prompting Ferrari to claim—risibly—that the car was an evolution of the SWB. Yes, it used a similar chassis and an upgraded version of the 3.0-liter V-12, now featuring a gleaming array of six Webers and producing 300 horsepower while spinning to 8,400 rpm. But Giotto Bizzarrini, who spearheaded the car's development, realized that the blunt bodywork of the SWB limited top speed to 155 miles per hour, so he focused on aerodynamics. The GTO's engine was fitted with a dry sump so it could be mounted lower in the chassis, and it was relocated behind the front axle to improve weight distribution. The Scaglietti bodywork was memorably lithe and muscular, with a long, low hood and a high rear deck ending in a spoiler that provided the downforce necessary at 175 miles per hour. When the GTO retired from front-line competition at the end of 1964, the sun set on a golden era of sports car racing.

DESIGNED FOR WINNING
1975 FERRARI 312 T

FUN FACT

Lauda closed out the 1975 season with a flag-to-flag win at Watkins Glen. But his teammate, Clay Regazzoni, was black-flagged for blocking Emerson Fittipaldi to allow Lauda to extend his lead.

HISTORICAL TIDBIT

The first US Grand Prix West at Long Beach in 1976 was the last win for the 312 T, as Regazzoni led Lauda to a crushing 1–2 finish.

KEY PERSON

Lauda claimed his second championship with the improved 312 T2 in 1977. Two years later, Jody Scheckter earned his only title with the further upgraded 312 T4. But the last variation on the theme, the ugly duckling 312 T5, was an epic fail.

It's always tempting to attribute success to a single great man or a paradigm shift in technology. But more often than not, it's the product of a confluence of factors large and small. So it was for Ferrari in 1974, coming off the worst stretch in its Formula 1 history. First, Mauro Forghieri, the mastermind behind the magnificent flat-twelve, was given control of F1 development. Then, Luca di Montezemolo, a brilliant lawyer being groomed by the Agnelli family for big things at Fiat, was brought in to run the race team. Montezemolo pragmatically axed the prototype sports car program to concentrate on F1, and a young, hard-headed, technically savvy Austrian named Niki Lauda was hired as the lead driver. Lauda and Forghieri tested incessantly at Ferrari's new private circuit at Fiorano. The fruit of their labor was the 312 T—T for *trasversale*. Forghieri chose to mount the transaxle transversely rather than longitudinally to concentrate the car's mass between the front and rear axle lines. Engineer Franco Rocchi tuned the engine to pump out 495 horsepower, which made it more than a match for the ubiquitous Cosworth DFV. The car looked every bit as impressive as it sounded, with a front wing that seemed to hover in midair and a huge cold-air box looming over the cockpit like a giant periscope. Lauda won four of five midseason races while breezing to the World Championship, and the car posted five consecutive victories at the end of 1975 and the beginning of 1976 before being supplanted by a slightly better iteration, the 312 T2. Ferrari earned four constructors' titles in five years before the discovery of ground effects rendered wide, flat engines like the horizontally opposed twelve-cylinder obsolete.

DESIGNED FOR WINNING
1994 FERRARI 333 SP

Enzo Ferrari dismissively referred to British Formula 1 constructors as *garagistas* because they built only chassis rather than entire cars. Not long after Enzo's death, Ferrari took a page out of the garagista playbook with the 333 SP, the marque's first purebred sports prototype since the 312 PB was mothballed in 1973. The impetus for the project came from Gianpiero Moretti, the debonair, cigar-smoking founder of the Momo automobile accessories and racing gear empire. Moretti's career had been launched when John Surtees put a Momo steering wheel in his championship-winning Ferrari 158 F1 car. Moretti competed as a gentleman driver in a series of flamboyant red-and-yellow prototypes, and he repeatedly begged Ferrari to return to sports car racing. The 333 SP—known as *Il Sogno Americano*, "The American Dream," because it was created to compete in IMSA's new World Sports Car class in the United States—was approved in 1993. To ensure that development of the new car didn't shortchange the F1 program, design and construction were shared with Dallara, another successful Italian race car manufacturer. Both the carbon fiber tub and the 4.0-liter, 650-horsepower engine were derived from Ferrari F1 technology. But the car was earmarked for sale to privateers, and no factory effort was ever contemplated. Four 333 SPs broke cover at Road Atlanta in 1994 and finished 1-2-5. The next year, they won the IMSA championship. Optimized for sprint racing in the States, the car was an also-ran at Le Mans. Still, a 333 SP scored a historic double at Daytona and Sebring in 1998, with Moretti co-driving. The car was no longer competitive by the time it was put out to pasture in 2003. But its rakish looks and the spine-tingling snarl of its V-12 never went out of style.

FERRARI 81

DESIGNED FOR WINNING
2002 FERRARI F2002

FUN FACT

Ferrari was so concerned about the weight of the F2002 that the team switched from cast-titanium to metal-matrix uprights, which saved about 1.75 pounds.

HISTORICAL TIDBIT

At the start at Monaco, Schumacher got bottled up behind Juan Pablo Montoya. Otherwise, he likely would have passed eventual race-winner David Coulthard and given the F2002 a perfect record for the season.

KEY PERSON

After dominating the Austrian Grand Prix, Barrichello was controversially ordered to allow Schumacher to win the race. Ferrari was fined $1 million for instituting team orders. Later, at Indianapolis, Schumacher returned the favor and handed the US Grand Prix to Barrichello.

The greatest Formula 1 car ever? You could make a strong case for the Ferrari F2002, the winner of fourteen of the fifteen races it contested in 2002. (It finished second, by barely a second, in the only one that got away.) The F2002 was the greatest monument to the Four Horsemen—driver Michael Schumacher, technical director Ross Brawn, designer Rory Byrne, and team manager Jean Todt—who had resurrected Ferrari's flagging fortunes in the mid-1990s. By 2002, they were coming off three consecutive constructors' championships, so they'd already amassed an impressive collection of greatest hits. But the F2002 was a quiet revolution that was immediately copied by just about every other team in the paddock. Lightweight components such as a fused titanium gearbox case allowed the design team to place ballast where it worked best, lowering the center of gravity. The drooping nose with the raised underside optimized airflow at the front of the car, while the waisted-in, Coke bottle–shaped body and unusually low side pods directed turbulent air away from the rear wing. The 3.0-liter V-10, credited to Paolo Martinelli and Gilles Simon, made 835 horsepower at 17,800 rpm, down slightly on the BMW, but it was implausibly compact and virtually indestructible. If the car had a fault, it was that it arrived late, three races into the season. But after that, Schumacher and Rubens Barrichello made mincemeat of the competition (such as it was). Nine races ended with 1–2 finishes. Schumacher clinched the driver's title with six races left. Together, he and Barrichello scored more points than the rest of the field combined. Thanks to the F2002, it was an F1 smackdown for the ages.

DESIGNED FOR WINNING
2007 FERRARI F2007

The mass exodus of the Holy Trinity—Michael Schumacher, Ross Brawn, and Rory Byrne—after the 2006 season ended the longest era of sustained dominance in Formula 1 history. The question everybody asked was whether Scuderia Ferrari could continue to thrive without its longtime superstars. The answer was a resounding *hell, yes*, thanks to the F2007. The 19,000-rpm, 2.4-liter V-8 was largely unchanged from the previous year, but the chassis had been completely overhauled. To improve airflow, the front end was lengthened significantly while the rear bodywork wrapped tightly around a sculpted carbon fiber gearbox case. The car also featured cutting-edge touches such as adjustable brake bias and a flexible floor that was later banned. In 2007, for the first time, Bridgestone was Formula 1's sole tire supplier, and it was widely believed that Ferrari had an advantage due to the team's longstanding relationship with the Japanese company. New boy Kimi Räikkönen won on his—and the car's—debut at Australia, and his teammate, Felipe Massa, scored back-to-back wins in the third and fourth races. But the season then turned into a yearlong knockdown-dragout battle with the McLarens driven by Fernando Alonso and Lewis Hamilton (who sometimes seemed to be fighting more with each other than with the Ferraris). Räikkönen, who was third going into the season-ending Brazilian Grand Prix, swept the last two races to beat Alonso and Hamilton to the championship by a single point. But the biggest news of the year turned out to be the "Spygate" scandal involving Ferrari mechanic Nigel Stepney, who was convicting of sharing technical secrets with McLaren. As a result, McLaren was stripped of its points, and Ferrari claimed the constructors' title.

DESIGNED FOR WINNING
2011 FERRARI 458 ITALIA GT

FUN FACT

The first major win for the 458 Italia was scored at Le Mans in 2012 by Toni Vilander, Gianmaria Bruni, and Giancarlo Fisichella—fitting, as these turned out to be the three most successful drivers in 458 history.

HISTORICAL TIDBIT

The success of a race car can be measured in part by how many teams choose to buy and race it. There were no fewer than fourteen 458s at Le Mans in 2014—and nineteen at Spa later that year.

KEY PERSON

In the States, the 458 flag was carried most proudly by Risi Competizione in Houston, Texas. Founder Giuseppe Risi owns the first dealership to be awarded after Ferrari North America was created in 1979.

A funny thing happened to sports car racing over the past decade. As prototypes—the big dogs at the top of the hill—grew ever more expensive and increasingly less relevant, manufacturers focused their attention on the lesser GT categories. Although GT cars don't compete for overall wins, they're immediately recognizable (think Corvettes or 911s) and race in large, entertaining packs. During the early 2000s, Ferrari was moderately successful with race versions of the F430. But the debut in 2010 of the 458 Italia gave the company a stellar sports car with a 570-horsepower V-8 in a state-of-the-art midengine chassis sheathed in aerodynamically efficient bodywork—a platform versatile and formidable enough to compete with anything in the GT classes. Ferrari collaborated with Michelotto Automobili, which had been involved in building the 333 SP prototypes in the 1990s, to develop a full line of 458s Italias—GT2 and GT3 models for Europe and the rest of the world and a Grand Am version specifically for the United States. The engine, bodywork, and mechanical specifications differed—radically, in some cases—according to the race series. But all the models shared one trait in common: they were race winners. Between 2011 and 2016, when the race version of the twin-turbo 488 GTB came online, 458 Italias scored more than one hundred outright victories and another hundred class wins. Although there were no factory entries, AF Corse, based in Piacenza, Italy, operated as the quasi-works team in international races. The 458 earned back-to-back-to-back World Endurance Championship manufacturers' titles, scored class wins at Le Mans in 2012 and 2014, and won the GT3 pro class in the 24 Hours of Spa five consecutive times.

FERRARI CHALLENGE

The genesis of Ferrari Challenge, the longest-running, production-based, single-make race series in the world, is a tale of transforming lemons into lemonade. Back in 1993, unpopular 348s were languishing unsold at entry ports and in showrooms. This led a marketing genius in Maranello to come up with the bright idea of creating a race series open only to—yep, you got it—348s modified with roll cages, safety gear, and a handful of performance tweaks. The concept was such a hit that, two years later, a Ferrari Challenge kit priced at $30,000 (installed by a dealer) was offered for the F355. That went over even better than the 348. Since then, the factory has sold Ferrari Challenge versions of the 360 Modena, F430, 458 Italia, and now the 488 GTB, which is so wicked that it lapped the Fiorano test track four seconds faster than the LaFerrari hypercar. Three race series are held in Europe, Asia, and North America, with a joint year-end bacchanal to crown international champions. Ferrari Challenge often runs as a support race at professional events. And the races themselves look like the real deal, with professionally built race cars, professional race teams, and professional drivers. But the paid shoes don't race themselves; they serve as coaches to the gentleman drivers who own the cars, and no ringers are permitted on the track. The series isn't designed to prepare young guns for the next level of competition. On the contrary, the idea is to promote the brand by coddling existing owners in a relaxed, swanky environment that encourages them to buy ever more, and ever-more-expensive, Ferraris. These days, similar high-end, one-make series are run by Porsche, Lamborghini, and Maserati. But Ferrari established the template that the other automakers follow.

GLOSSARY

CONSTRUCTORS' CHAMPIONSHIPS: Fans and the media tend to focus on the drivers, so attention naturally is riveted on the drivers' titles. But in Formula 1 and most forms of sports car racing, a second championship is awarded to the manufacturers, or constructors, whose cars amass the most points during a season.

DRY SUMP: Most street cars have a wet sump lubrication system. Oil sits in a pan, or sump, beneath the crankcase. A single pump routes oil to the engine, and the fluid drains back into the pan. A dry sump, found in many exotic cars and race cars, uses a remote reservoir to hold the oil and two pumps to get the fluid to and from the engine. A dry sump allows the engine to sit lower in the chassis and provides better lubrication under high cornering loads.

HOMOLOGATION: In certain classes of sports car racing, manufacturers must build a specified number of cars—twenty-five, fifty, one hundred, whatever—before the model can be homologated, or approved, for competition. The idea is to prevent manufacturers from creating purebred race cars (i.e., prototypes) for production-based categories.

INDEPENDENT SUSPENSION: The first automobiles were equipped with a pair of solid axles, so each set of wheels moved in tandem, with jarring results. Independent front suspensions, so called because they allow the front wheels to move independently, were commonplace by the late 1940s, but solid rear axles remained the norm until the 1960s. The first Ferrari street car with a fully independent suspension was the 275 GTB in 1964.

PROTOTYPES: In sport car racing, "prototypes" refer to purpose-built thoroughbreds that compete for overall victory in endurance classics like Le Mans and the Rolex 24. In the racing hierarchy, they stand above production-based sports cars, which race in the GT categories.

TRANSAXLE: A transaxle combines the functions of the transmission and the driven axle in a single unit. Practically speaking, it's a single component containing the gearbox and differential. Transaxles are standard in most front-wheel-drive and midengine cars. The Ferrari 275 GTB and Daytona were two of the rare front-engine, rear-wheel-drive cars fitted with transaxles.

THE HEROES
BEHIND THE WHEEL

THE HEROES BEHIND THE WHEEL
ALBERTO ASCARI

FUN FACT

In 1952, Ascari became the first—and last—driver to race a works Ferrari in the Indy 500. Three other privately owned Ferraris practiced that year, but Ascari was the only driver to make the show. His car broke forty laps into the race.

HISTORICAL TIDBIT

Ascari's contract with Lancia stipulated that he didn't have to compete in the ultra-dangerous Mille Miglia. When his mentor, Luigi Villoresi, was injured during practice in 1954, he reluctantly agreed to race—and won in a Lancia D24.

KEY PERSON

Like his father, Ascari died at age thirty-six on the twenty-sixth day of the month, in an ambulance speeding to a hospital. Supremely superstitious, Ascari wasn't wearing his lucky light-blue helmet when he was killed.

No driver looms larger in Ferrari lore than Alberto Ascari. Not Lauda. Not Villeneuve. Not even Schumacher. The record book shows that Ascari won Ferrari's first two World Championships. But more than that, he was the driver whose results made the marque a go-to destination for decades to come. The son of the celebrated Antonio Ascari, he was only seven years old when his father was killed while leading the French Grand Prix in 1925. Ascari defied his mother's wishes and started racing motorcycles as a teenager, and when he made his four-wheel racing debut in the1940 Mille Miglia, he was driving Enzo Ferrari's first race car, the Auto Avio Costruzioni 815. After the war, he became a works driver for Ferrari and established himself, like his father, as the premier Italian racer of his generation. Only a poor tire choice in the last race of the season prevented him from beating Juan Manuel Fangio and winning the Formula 1 title in 1951. The next year, after missing the season opener, he ran the table. The year after that, he won five of eight starts to claim his second consecutive championship. Only Fangio was in his class, and when Ascari was at his best, he was close to unbeatable. He quit Ferrari to join Lancia after Enzo refused to meet his salary demands. Ascari was leading at Monaco in 1955 when he overcooked the chicane and careened into the harbor. Four days later, while wearing street clothes, he impetuously climbed into a friend's car during a test at Monza and crashed fatally at what's now known as the Variante Ascari. The car he was driving? A Ferrari 750 Monza.

THE HEROES BEHIND THE WHEEL
MIKE HAWTHORN

FUN FACT

Hawthorn was right behind Collins—whom he called "*mon ami*, mate"—when his friend crashed and died at Pflanzgarten. One month earlier, in his side-view mirror, he'd seen another Ferrari teammate, Luigi Musso, killed in the French Grand Prix.

HISTORICAL TIDBIT

Junior teammate Phil Hill pulled over to let Hawthorn past near the end of the last two Grand Prix races of the 1958. The extra points that Hawthorn earned gave him the F1 title.

KEY PERSON

Hawthorn was disqualified for pushing his car against the direction of traffic at the Portuguese Grand Prix in 1958. But he was reinstated to second place after Moss sportingly testified on his behalf.

With his blond hair, bow ties, and tweed jackets, Mike Hawthorn looked like the most English of drivers. He was intensely patriotic, and he desperately wanted to drive a British car into victory lane. It's a cruel irony, then, that his greatest triumphs were earned in Ferraris while tragedies befell him in his beloved Jaguars. Hawthorn's eat-drink-and-be-merry-for-tomorrow-we-die ethos made him the polar opposite of his more professional country-man—and principal rival—Stirling Moss. The son of a garage owner who specialized in high-performance cars, Hawthorn came to Ferrari's attention while slaying giants in an underpowered Cooper-Bristol. In 1953, he was signed at the age of twenty-four to race the all-conquering Ferrari 500. During his first few races, he was outclassed by his more experienced team-mates. Then, in the French Grand Prix, he shockingly out-dueled Juan Manuel Fangio to win what was immediately dubbed "the race of the century." Hawthorn later left Ferrari in a quixotic quest to win a Formula 1 race in a British car. But his only subsequent successes came in sport cars, and even there he was unlucky. His victory at Le Mans in a Jaguar D-type was marred by accusations that he caused the catastrophic accident that killed more than eighty people. Hawthorn eventually returned to Ferrari, and, despite winning only one race, he beat Moss by a single point to earn the World Championship in 1958. But it was a hollow victory: his close friend, Peter Collins, was killed when his Ferrari rolled at the Nordschleife. Heart-broken, Hawthorn quit racing at the end of the year. Three months after retiring, he died when he wrecked his Jaguar sedan on a wet highway near his home in England.

THE HEROES BEHIND THE WHEEL
PHIL HILL

FUN FACT

Hill was a rare driver who won his first and last races—in a clunky MG TC at Carrell Speedway in Los Angeles in 1949 and in a high-wing Chaparral 2F, the most advanced car of its day, in the BOAC 500 at Brands Hatch in 1967.

HISTORICAL TIDBIT

After retiring from racing, Hill and Ken Vaughn formed Hill & Vaughn, which was among the first high-end restoration shops in the United States. The company was responsible for several cars that won awards at the Pebble Beach Concours d'Elegance.

KEY PERSON

Hill's son, Derek, was a racing driver whose first notable successes came in the Ferrari Challenge series, driving 348s and then F355s, before transitioning into formula car racing.

Growing up in Southern California, Phil Hill belonged to the Santa Monica Low Flyers, a club whose members included several future hot rod legends. But Hill gravitated to MGs rather than flathead Fords, and he became one of the first Americans to venture to Europe in compete in big-time road racing. Although he was notoriously high-strung outside the cockpit, his smooth driving technique, technical knowledge, and mechanical sympathy made him a long-distance racer par excellence. He won Le Mans for Ferrari for the first time (of three) in 1958 after grueling nighttime stints in torrential rain, and he would go on to win the 12 Hours of Sebring three times and the 1000 Kilometers of the Nürburgring twice. His first time out in a Formula 1 Ferrari, he led early at Monza and set fastest lap. In 1960, at the same track, he became the first American to win a Grand Prix since Jimmy Murphy won the French Grand Prix in a Duesenberg in 1921. In 1961, with Ferrari holding the upper hand with its shark-nose 156, the championship fight came down to an inter-team battle between Hill and German nobleman Wolfgang von Trips. In the second-to-last race of the season, again at Monza, von Trips touched wheels with Jim Clark and crashed fatally at the Parabolica, killing fifteen spectators. Hill cruised to victory and clinched a bittersweet World Championship. He quit Scuderia Ferrari the next year after mass defections undermined the once-powerful team—a mistake, as it turned out, that ended his career as a competitive F1 driver. But he continued to race against Ferrari in Ford GT and Chaparral prototypes, and he's now remembered as one of the finest gentlemen that motorsports ever produced.

THE HEROES BEHIND THE WHEEL
JOHN SURTEES

FUN FACT

Thanks to exhaust fumes that leaked into the cockpit of his 250 P, Surtees "celebrated" his win at Sebring in 1963 in his debut for Ferrari by vomiting explosively for ten minutes after the race.

HISTORICAL TIDBIT

For political reasons, Ferrari didn't officially contest the Grands Prix at Watkins Glen and Mexico City in 1964. The cars were therefore entered by Luigi Chinetti's North American Racing team, and they carried white-and-blue livery for both races.

KEY PERSON

Surtees owned a pair of 330 GT 2+2 road cars during his tenure at Ferrari, putting about thirty thousand miles on each of them. He sold one to idiosyncratic *Road & Track* writer Henry Manney and the other to Chinetti.

John Surtees was the first man to win World Championships on two wheels and four. He's almost certainly the last. "Il Grande John," as he was fondly known in Italy, became a legend for his fearless rides on MV Agustas. In 1963, with only three years of car racing experience, he was summoned to Maranello for an audience with Enzo Ferrari. Surtees was hired to help Ferrari recover from the failed mutiny known as "the Purge." Driving a 156 Formula 1 car without its signature shark nose, Surtees scored an epic win at the Nordschleife in 1963. The next year, in a 158 powered by a formidable V-8, he fought a season-long battle with Jim Clark in a Lotus and Graham Hill in a BRM. By finishing second in Mexico at the last race of the year, sneaking past his teammate, Lorenzo Bandini, shortly before the checkered flag, Surtees scored just enough points to claim the championship. In those days, drivers were expected to race sports cars as well as F1, and Surtees was the acknowledged team leader in prototype racing, winning Sebring once and the Nürburgring twice. But he had two strikes against him: First, he wasn't Italian. Second, team manager Eugenio Dragoni was determined to develop homegrown talent into driving aces. The political tension peaked at Le Mans in 1966, where Ferrari was facing an armada of Ford GTs. When Dragoni nominated Ludovico Scarfiotti to start the race, Surtees sped to Modena to confront Ferrari. Enzo refused to countermand Dragoni's order, so Surtees quit. Fords finished 1-2-3, and Ferrari never again won Le Mans. Surtees scored his last Grand Prix victory in a Honda—at Monza, fittingly—with no Ferrari even on the lead lap.

THE HEROES BEHIND THE WHEEL
NIKI LAUDA

FUN FACT

Lauda won the Swedish Grand Prix in 1978 in the Brabham BT46B, popularly known as the "fan car," designed by Gordon Murray with a novel form of ground effects that sucked the car to the ground. The one-race wonder was effectively banned and never competed again.

HISTORICAL TIDBIT

An avid private pilot, Lauda ran his own commercial airline, Lauda Air, for more than a decade before selling the company to Austrian Air. He later founded another commercial airline, Niki, but left it to return to F1.

KEY PERSON

Notwithstanding the dramatics in the movie *Rush*, Lauda's chief rival wasn't British playboy James Hunt but the phlegmatic Frenchman Alain Prost, his teammate at McLaren, whom he beat to the title in 1984 by a mere half point.

Niki Lauda struck most observers as a strange hire when he joined Scuderia Ferrari in 1974. He was a rich Austrian who'd bought his way into Formula 1, and he was best known for his headstrong personality and pronounced overbite. Although his handsome, exuberant teammate, Clay Regazzoni, was the favorite of the tifosi, the dispassionate and more mechanically savvy Lauda bonded with technical director Mauro Forghieri and team manager Luca di Montezemolo to revive a team that was dead in the water. Lauda led Regga to a pair of 1-2 finishes in 1974 and lost the championship only through a lack of experience and mechanical unreliability. But Forghieri's new 312 T—"T" for the transverse-mounted gearbox—sat at the top of the F1 totem pole. Lauda breezed to the title in 1975 and won five of the first nine races in 1976 before crashing at the Nordschleife. His Ferrari hurtled into a rockface, burst into flames, and was speared by an oncoming car. By the time he was yanked from the cockpit, Lauda had inhaled toxic fumes, and his scalp and right ear were hideously burned. He lingered near death for four days. Yet a mere six weeks later, with his scalp still oozing blood, he mastered his fears and finished fourth at Monza. James Hunt claimed the championship when Lauda, convinced that conditions were unsafe, withdrew from the season-ending race in Japan. Enzo Ferrari never forgave him, and, even though Lauda won his second title the next year, the two parted ways acrimoniously. Lauda went on to win a third World Championship for McLaren. He's now a minority owner of the Mercedes team that is Scuderia Ferrari's greatest nemesis.

THE HEROES BEHIND THE WHEEL
GILLES VILLENEUVE

FUN FACT

Villeneuve's son, Jacques, who was ten years old when his father died, won the Indianapolis 500 in 1995 and the World Championship in 1997. But he never won the Canadian Grand Prix—held at Circuit Gilles Villeneuve.

HISTORICAL TIDBIT

Villeneuve learned to drive in slippery conditions by racing snowmobiles. When he attended his first race car driving school, the chief instructor called the owner to breathlessly inform him that Villeneuve was lapping faster than he was.

KEY PERSON

Pironi's career was vexed after Villeneuve's death. Riccardo Paletti was killed when he ran into Pironi's stalled Ferrari in Canada. Then, while handily leading the championship, Pironi sustained career-ending injuries in a crash in Germany. He died in 1987 in an offshore powerboat race.

Gilles Villeneuve never won a World Championship, and he won only six Formula 1 races. In fact, he's probably best known for the ones he *lost*—wheelie-ing gloriously (if pointlessly) around Zandvoort after crashing and flattening a tire or off-roading and banging wheels with Rene Arnoux to finish second at Dijon. Enzo Ferrari, who was notoriously callous toward his drivers, adored Villeneuve like a surrogate son, so it's by providential coincidence that the plaque memorializing him near the test track at Fiorano is a block from Via Dino Ferrari. Villeneuve was born with supernatural car control and no apparent capacity for fear. This combination allowed him to perform feats that left his rivals slack jawed. The US Grand Prix at Watkins Glen in 1979 is remembered less for his majestic victory than for the exhibition he put on in a deluge during practice, when he was nearly ten seconds—ten seconds!—faster than his teammate, Jody Scheckter, who'd just clinched the World Championship. Villeneuve scored his first Grand Prix win as though it had been scripted, in Montreal, in front of delirious fellow French-Canadians. His surprise victory in Spain in 1981, in the powerful but sled-like 126 CK, was one of the greatest defensive drives ever. Fair to a fault, he scrupulously followed team orders and allowed Scheckter to claim a title that might have been Villeneuve's if he'd been more selfish. So, he felt especially betrayed when, in his opinion, teammate Didier Pironi reneged on a pre-race deal and won the Grand Prix of San Marino in 1982. Two weeks later, at Zolder, Villeneuve misjudged a risky passing maneuver during qualifying and died after clipping a slower car. Scheckter eulogized him as "the fastest racing driver history has ever known."

THE HEROES BEHIND THE WHEEL
MICHAEL SCHUMACHER

Ayrton Senna created the paradigm for the modern race car driver, but Michael Schumacher perfected it. Technically adept, physically fit, and utterly ruthless both on and off the track—these were the attributes that Schumacher employed to redefine what was expected of drivers not only in Formula 1 but all the way down to go-karting. It was only appropriate that, after Senna was killed at Imola in 1994, Schumacher went on to win the race as well as his first World Championship. In 1996, he joined a Ferrari team mired in mediocrity. With his never-say-die driving and remarkable team-building skills, Schumacher transformed Scuderia Ferrari into a contender, even though the F310 was no match for the more sophisticated Williams FW18. In 2000, Schumacher began the most dominant stretch in Grand Prix history, winning the first of five consecutive championships. The victories didn't come without controversy. In 1997, he was excluded from the year-end results because he tried to wreck Jacques Villeneuve at Jerez in a last-ditch effort to win the title, and in 2006, he was ignominiously demoted to the back of the grid for "crashing" intentionally at Monaco to end the qualifying session prematurely. Schumacher also got a lot of flak for insisting on Number 1 driver status and reducing his teammates to second-class citizens. The public outrage after Rubens Barrichello was told to pull over to allow Schumacher to win the Grand Prix of Austria in 2002 compelled the FIA to ban team orders. Schumacher won his final race in China in 2006 in wet-dry conditions that showcased his superior car control and grasp of tactics—a capstone to a career of uniquely sustained excellence.

THE HEROES BEHIND THE WHEEL
SEBASTIAN VETTEL

FUN FACT

Vettel's childhood heroes were "the three Michaels"—Michael Schumacher, Michael Jordan, and Michael Jackson. His singing aspirations evaporated after puberty caused his voice to change.

HISTORICAL TIDBIT

Vettel was nineteen years, eleven months old when he made his F1 debut at the US Grand Prix in 2007. After a poor start, he clawed his way back to eighth place and became the youngest driver to score a championship point.

KEY PERSON

Vettel's win at Monaco in 2017 was the tenth for Ferrari on the streets of Monte Carlo. The first was scored in 1955 by mustachioed Frenchman Maurice Trintignant, who'd won Le Mans the previous year in a mighty Ferrari 375 Plus.

It's not fair to refer to Sebastian Vettel as Michael Schumacher 2.0, because that would imply he's a new-and-improved version of his idol. And the jury's still out on that score. But there's no mistaking the uncanny parallels between the two Germans. Both were racing go-karts by the time they were eight. Both were wunderkinds who rocketed through the lower formulas. Both made their Formula 1 debut as last-minute replacements, and both started seventh in their first F1 outing. Vettel was even more precocious than Schumacher, having joined the Red Bull Junior Team when he was only eleven, and he was an F1 regular by the time he was twenty. In 2008, he became the youngest driver to win a Grand Prix when he scored a majestic victory in the rain at Monza in a Toro Rosso powered by a Ferrari engine. Vettel later won four consecutive championships at Red Bull, breaking the hearts of the tifosi while denying Fernando Alonso the titles for Ferrari in 2010 and 2012. Vettel replaced Alonso at Ferrari in 2015 and notched an emotional victory at Malaysia in his second race for the team. Like Schumacher, Vettel has been overcome by the red mist on several occasions, from the notorious "Multi 21" incident in Malaysia in 2013, where he ignored team orders and overtook Red Bull teammate Mark Webber, to purposely banging wheels with Lewis Hamilton's Mercedes during the Azerbaijan Grand Prix in 2017, which earned him a "dangerous driving" penalty. But, again like Schumacher, Vettel has done yeoman work leading a Ferrari renaissance, and the thoughtful German seems likely to remain the face of the F1 program for years to come.

THE HEROES BEHIND THE WHEEL
FANGIO, SCHECKTER & RÄIKKÖNEN

FUN FACT

Räikkönen moved from third to first in the championship hunt in the final race of 2007. John Surtees performed the same feat in 1964 in his Ferrari 158.

HISTORICAL TIDBIT

Six drivers scored their only World Championship victories in a Ferrari: Piero Taruffi, Luigi Musso, Giancarlo Baghetti, Lorenzo Bandini, Ludovico Scarfiotti, and Jean Alesi (who had trouble seeing during the final laps—at Circuit Gilles Villeneuve, coincidentally—because he was crying inside his helmet).

KEY PERSON

Mario Andretti won the South African Grand Prix in 1971 in his first race for Ferrari. He drove his last F1 race—in a Ferrari—in Las Vegas eleven years later.

Three drivers—Juan Manuel Fangio, Jody Scheckter, and Kimi Räikkönen—have won World Championships for Ferrari without ever seeming to fully embrace the Ferrari mystique or to be fully embraced by the Ferrari family. First and foremost was Fangio, who spent most of his career fighting against Ferrari, first with Alfa Romeo and then with Maserati. In 1956, coming off back-to-back championships with Mercedes-Benz, he joined Ferrari only because the German manufacturer had pulled out of racing. Even so, he wouldn't have earned the title if Peter Collins hadn't chivalrously allowed him to take over his car in the middle of the Italian Grand Prix and score the points he needed to beat Stirling Moss. Nevertheless, Fangio left Ferrari the next year and won his fifth and final championship for Maserati. Scheckter, too, needed help to claim his title for Ferrari in 1979. Although he'd come into Formula 1 with a reputation as the fastest, most spectacular driver around, that title had been usurped by his young teammate, Gilles Villeneuve. Villeneuve was generally quicker than Scheckter, but he deferred to his team leader and finished second in a championship 1-2. Scheckter suffered through a thoroughly miserable season the next year and promptly retired. Räikkönen, meanwhile, is the most un-Ferrari-like driver imaginable. A taciturn Finn whose deadpan, monosyllabic public persona has earned him the not entirely flattering moniker "The Iceman," Räikkönen seemed like an odd choice when he was poached from McLaren in 2007 to replace the retiring Michael Schumacher. But he won the World Championship his first year with Ferrari, barely beating McLaren's Lewis Hamilton and Fernando Alonso. He's now in the middle of his second tour of duty with the team.

THE HEROES BEHIND THE WHEEL
MASSA, PROST & ALONSO

Nine men have won World Championships for Ferrari. More than a dozen finished second in the title chase, in some cases losing by no more a single point. None suffered a crueler fate than Felipe Massa. The diminutive, baby-faced Brazilian spent eight years with Scuderia Ferrari, loyally supporting Michael Schumacher, then Kimi Räikkönen, and finally Fernando Alonso. But in 2008, he unexpectedly outpaced Räikkönen and ran neck and neck with McLaren's Lewis Hamilton. At his home track in Sao Paulo, during the last race of the season, Massa crossed the finish line first, briefly vaulting him into the championship lead. But even as Massa's family and fans rejoiced, Hamilton passed Timo Glock on the last corner of the last lap to finish fifth and eke out just enough points to snatch away the title. In 1990, Alain Prost saw his championship hopes evaporate at the first corner of the first lap in the Japanese Grand Prix, bringing down the curtain on the final act of his incendiary feud with Ayrton Senna. The previous year, while they were teammates at McLaren, Prost had secured his championship when he closed the door on Senna near the end of the race at Suzuka. The next year, with Prost now driving the Ferrari 641, Senna purposely speared Prost ten seconds after the start to secure *his* championship. Fast-forward twenty years to 2010, when double World Champion Fernando Alonso joined Ferrari. After leading the title chase for most of the year, Alonso got trapped behind Vitaly Petrov near the end of the Abu Dhabi Grand Prix and came up four measly points short of Sebastian Vettel. "Sometimes you win," Alonso said afterward. "Sometimes you lose."

SPORTS CAR ACES

Ferrari quit racing sports cars after failing to win a championship in 1973 and has focused on Formula 1 ever since. But the victories at Le Mans and the classic endurance races made the marque an international icon. So, it seems only fair to shower a little love on the lesser-known drivers who helped put Ferrari on the map. First on the list is Clemente Biondetti, a Sardinian whose prewar career was stalled by his anti-fascist views. In 1948, when he was nearly fifty years old, he gave Ferrari its first memorable wins in the Mille Miglia and the Targa Florio. The following year, against all odds (and Tazio Nuvolari), he repeated his improbable double. The most prominent sports car hero of the next generation was the wealthy Belgian Olivier Gendebien, who was usually paired with Phil Hill. Although Gendebien wasn't Hill's equal in single-seaters, he won Le Mans four times while clocking three victories each at Sebring, the Targa Florio, and the Tour de France. Later, during the fierce wars fought with the Ford GT, Ludovico Scarfiotti won at Sebring, Le Mans, the Nürburgring, and Spa-Francorchamps. Meanwhile, Nino Vaccarella triumphed at Le Mans, the 'Ring, and the Targo Florio. In 1970, at Sebring, he also shared the first and last major win for the 512 S—a car that was more famous for its exploits in the movie *Le Mans* than for its accomplishments in the real world. Ferrari made a soft return to prototype racing in the 1990s with the 333 SP. Didier Theys raced the car for six consecutive years, scoring an American double at Daytona and Sebring in 1998. For the past decade, Ferrari has been involved in GT racing with the F430, the 458 Italia, and now the 488 GTB, with Toni Vilander, Gianmaria Bruni, and Giancarlo Fisichella collecting dozens of class wins.

GLOSSARY

CHAMPIONSHIP POINTS: Formula 1 has used several systems to calculate points for the drivers' and constructors' championships. Initially, points were awarded from first to fifth place—8-6-4-3-2—with an additional point earned for the fastest lap. From 1961 to 1990, the scoring was extended to include sixth place: 9-6-4-3-2-1. Since 2010, points have been paid out to tenth place: 25-18-15-12-10-8-6-4-2-1.

CO-DRIVERS: In endurance racing, where races last as long as twenty-four hours, multiple drivers are necessary to compete. Well into the 1980s, many teams raced at Le Mans with only two drivers. Currently, three is the norm, and four isn't uncommon. During the early years of F1, one driver occasionally took over for another during a race. But points ceased being awarded for shared drives, and the practice ended in the 1960s.

FORMULA 1 CHAMPIONSHIPS: The World Championship title has been awarded to a driver every year since F1 was created in 1950. A second championship for constructors was instituted in 1958. Ferrari drivers have collected fifteen World Championships, more than any other constructor. (McLaren is second with twelve.) Ferrari has also won the most constructors' titles: sixteen. Williams is second with nine.

MONACO GRAND PRIX: While the French Grand Prix has been held at more than a dozen venues over the years, the Grand Prix of Monaco has been staged only on the streets of Monte Carlo, on a circuit that's changed remarkably little since the inaugural race in 1929. The first of Ferrari's nine Formula 1 wins at the principality was scored by Frenchman Maurice Trintignant in an ungainly Ferrari 625.

NON-CHAMPIONSHIP RACES: Nowadays, all F1 races count toward the year-end drivers' and constructors' championships. But in the 1950s and 1960s, numerous non-championship races drew large fields of F1 cars. Some, like the Race of Champions at Brands Hatch, were quite prestigious. The second-tier British Formula 1 Series, which folded in 1982, was the last venue for non-championship F1 racing other than the final Race of Champions in 1983.

NORDSCHLEIFE: The Nürburgring, the premier motorsports complex in Germany, was completed in 1927. Until the 1970s, major races were held on the Nordschleife, or "North Loop." Nicknamed "The Green Hell," the Nordschleife consisted of eighty-nine left-hand turns and eight-four right-handers on a 14.2-mile-long circuit. F1 stopped racing there after Niki Lauda's fiery wreck in 1976, and the German Grand Prix is now run on a shorter modern circuit at the Nürburgring. But the Nordschleife is still used for track days and sports car races.

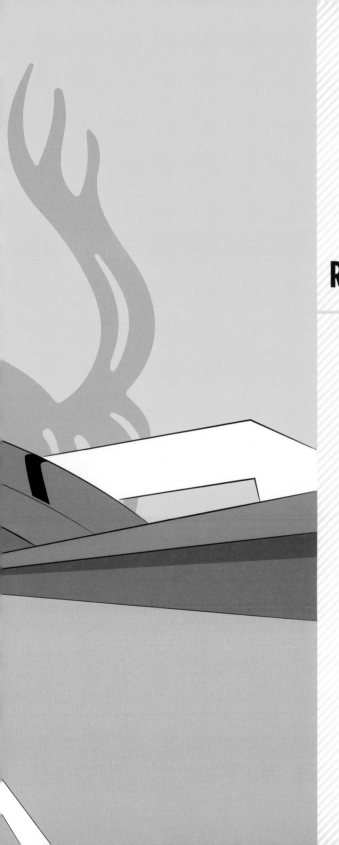

RACES TO REMEMBER

FUN FACT

The Chinetti/Selsdon car was privately entered, so the win in the 1954—with José Froilán González and Maurice Trintignant at the wheel of a brutish 375 Plus—was Ferrari's first at Le Mans.

HISTORICAL TIDBIT

Chinetti's iron-man performance at Le Mans in 1949 came in stark contrast to his win in 1932, when he was sick, and his teammate, Raymond Sommer, drove all but three or four hours.

KEY PERSON

Three weeks after Le Mans, Chinetti drove a sister 166 MM to victory in 24 Hours of Spa despite crashing into a house, injuring a woman, and stopping to administer first aid.

Since 1923, the 24 Hours of Le Mans has been the racing world's most punishing and prestigious test of automotive endurance, and carmakers such as Bentley, Jaguar, Ford, Porsche, and Audi have used it to polish their brand image. Two Ferrari 166 MMs were entered in the first postwar event in 1949. Although the cars had already won a few races in Italy, they didn't generate much buzz in France, and most of the pre-race coverage focused on the Talbot-Lagos, Delages, and Delahayes. The second Ferrari wasn't a factor in the race. But the other car, driven by Luigi Chinetti, an old friend of Enzo Ferrari's, ran with the leaders from the start. Chinetti had worked with Scuderia Ferrari during the prewar years, and he'd won Le Mans twice in Alfa Romeos. By midnight, he'd pushed his pretty little Barchetta to the top of the leaderboard. At 4:26 a.m., just past the halfway point, he pitted to turn the Ferrari over to the car's owner, Peter Mitchell-Thomson, a.k.a. Lord Selsdon. For reasons that are cloudy, Lord Selsdon drove only one short stint that lasted, according to conflicting reports, either twenty minutes or seventy-two minutes. Chinetti, nearly forty-eight years old at the time, was understandably exhausted as the race wound down. Making matters even worse, the cockpit was slick with oil and the clutch started slipping. The driver of the second-place Delage sped up in a banzai effort to catch the Ferrari. But Chinetti held on to win by a lap. The 2.0-liter V-12 in the Ferrari was the tiniest engine to win Le Mans until a Porsche 919 Hybrid triumphed with a marginally smaller turbocharged powerplant in 2015.

"I have killed my mother!" A confession of matricide? No, just Enzo Ferrari's poignant—if characteristically theatrical—admission after the British Grand Prix in 1951. Since going into business, Ferrari had been obsessed with the dream of humbling his old employer, Alfa Romeo, which had ruled postwar Grand Prix racing with a supercharged, straight-8 prewar beauty dubbed the 158/159. Ferrari had failed to beat the so-called Alfettas with the 1.5-liter supercharged Colombo V-12. Now, he was gambling on an un-supercharged 4.5-liter V-12 designed by Aurelio Lampredi. The Ferrari 375 had finished second to the Alfa in the first three races of the season. But here at Silverstone, on a flat, barren circuit fashioned out of a wartime airfield, José Froilán González astonished everybody by planting his Ferrari on the pole. As soon as the race began, he and Alfa Romeo's team leader, Juan Manuel Fangio, checked out from the field. Squat and portly, González appeared to be overflowing from the cockpit as he energetically slapped on armfuls of opposite lock. Whereas Fangio was a superb stylist, González power-slid his 375 around Silverstone's wide-open corners like a drift king, occasionally taking to the grass and once bouncing off the hay bales at Becketts. González and Fangio were still dicing furiously at half distance. But the more highly stressed Alfa was far thirstier than the Ferrari, and Fangio was never able to make up the time lost during his longer pit stop for fuel. González won by fifty-one seconds. Ferrari sent a telegram to Alfa Romeo: "I still feel for our Alfa the adolescent tenderness of first love." Tenderness, yes, but also the satisfaction of revenge long delayed.

1953 FRENCH GRAND PRIX

FUN FACT

No Ferraris raced at Reims in 1962 because of a strike in Italy. When the team next won again in the French GP, it was at another circuit.

HISTORICAL TIDBIT

With its long straights, Reims favored cars with lots of power, so Ferraris traditionally fared well there. Tony Brooks led Phil Hill to a Ferrari 1-2 in 1959, and Giancarlo Baghetti scored his only World Championship victory at Reims in 1961.

KEY PERSON

Hawthorn won his third—and last—Grand Prix during his final visit to Reims in 1958. The race was marred by the death of his teammate Luigi Musso, who crashed while chasing Hawthorn through the scary-fast right-hander past the pits.

Mike Hawthorn was a long shot going into the 1953 French Grand Prix. The twenty-three-year-old Briton was still finding his bearings in Scuderia Ferrari, while team leader Alberto Ascari hadn't lost a World Championship race in the Tipo 500 since, well, ever. (He was nine for nine.) Also starting ahead of Hawthorn were two former World Champions—Juan Manuel Fangio and Nino Farina—and the terrifically fast José Froilán González. Reims was a roughly triangular circuit consisting mostly of long straights broken up by slow corners. González, starting with a half tank of fuel, sped off into the distance. But behind him, the race settled into a fierce slipstreaming contest between Hawthorn, Fangio, Ascari, and his mentor, Gigi Villoresi. "At one point I passed Ascari, and he shrugged his shoulders as if to say, 'Take it away; I can't go any faster!'" Hawthorn wrote later. "Positions were changing several times a lap. I had the lead and then Villoresi came past, and sometimes we would be hurtling along, three abreast at 160 miles per hour, down an ordinary French main road." The lap record was broken seven times by four drivers. After González pitted, the race came down to Hawthorn and Fangio in his six-cylinder Maserati, the old man and the new boy, swapping the lead lap after lap. They crossed the finish line side by side on Lap 58. Next time by, Hawthorn led by a heartbeat. On the final lap, Fangio was in front approaching Thillois, the hairpin leading to the front straight. When his gearbox balked, Hawthorn scythed past under braking. Hawthorn held on to win by less than a second and broke down in tears when the British national anthem was played.

FUN FACT

The differential on the winning car failed on the way back to the paddock after the finish. Neither Rindt nor Gregory ever completed another race at Le Mans.

HISTORICAL TIDBIT

Scuderia Ferrari team manager Eugenio Dragoni, ostensibly speaking for Enzo Ferrari, pleaded with Chinetti to allow the second-place car—which was fitted with Dunlop tires—to win the race. Chinetti, who had a sometimes tempestuous relationship with Enzo, refused to back off.

KEY PERSON

Although there are no official records or contemporary reports supporting his claim, reserve driver Ed Hugus—a Pittsburgh Ferrari dealer and friend of Chinetti's—says he drove a stint during the night because Gregory was too tired to continue and Rindt couldn't be found.

The 24 Hours of Le Mans in 1965 was supposed to be a clash of titans—Ford versus Ferrari, the corporate might of America against the artisanal genius of Italy. It turned to be more like a street fight between spear carriers, and the guys who won would have been voted "most unlikely to succeed" before the race. Ford started with eleven cars. Only one made it to the end, 373 miles behind the winner. Meanwhile, there were no fewer than twelve Ferraris, and at one point, they were running 1-2-3-4-5-6. But then they, too, started breaking—cracked rotors, disintegrating gearboxes, pulverized engines. Through his North American Racing Team, Luigi Chinetti had entered a Ferrari 250 LM for a pair of past and future Formula 1 drivers, Masten Gregory and Jochen Rindt. Because it had been designed for the GT category, the LM wasn't quick enough to hang with the prototypes, especially as it was riding on Goodyear rain tires that weren't designed for it. The car was bedeviled by mechanical issues early on, and, when Gregory pitted with what sounded like a terminally ill engine, he found Rindt in street clothes, ready to leave for his hotel. Fortunately, Rindt's rental car was blocked, and Gregory was able to cajole him into staying. The two of them agreed to drive flat out to the finish, turning Le Mans into a twenty-four-hour grand prix. Lo and behold, all the prototypes broke, and Gregory nursed a failing differential during the final stint to win by five laps over another privately entered 250 LM. None of the factory entries finished, and Ferrari never won Le Mans again.

RACES TO REMEMBER
1974 SPANISH GRAND PRIX

Scuderia Ferrari suffered through a terrible, horrible, no good, very bad Formula 1 season in 1973. The car was so badly outclassed by the Cosworth DFV-powered brigade that the team simply stopped showing up for races. Toward the end of the year, Mauro Forghieri was reinstalled as technical director. Clay Regazzoni was rehired to drive one car, while the other was entrusted to an unproven Austrian, Niki Lauda, whose confidence far outstripped his results to date. Lauda and Forghieri tested relentlessly during the off-season, and the revised, flat-twelve–powered 312 B3 showed promise early in 1974. Regazzoni was third in Argentina, then second in Brazil, and Lauda qualified on the pole in South Africa. All the pieces fell into place at Jarama. Lauda and Regazzoni started 1-3. Ronnie Peterson—an acknowledged rainmaster—led early on a wet track, but Lauda patiently stalked him as the pavement dried and thundered past on the front straight. Eventually, everybody stopped for slicks. While other teams fumbled their tire changes, Ferrari's fast, efficient pit stops kept Lauda and Regazzoni well in front. Then, it was simply a matter of cruising imperiously to the finish. Lauda set the fastest lap, and, even though the race was stopped prematurely, he took the checkered flag thirty-five seconds ahead of Regazzoni. No other car was on the lead lap. The rest of the season was marked by ups (another 1-2 at Zandvoort) and downs (several wrecks and mechanical failures). But the Spanish GP was the race that reestablished Ferrari as a forced to be reckoned with. Ten months later, the seminal 312 T debuted, and Scuderia Ferrari would be the team to beat for the rest of the decade.

Sporting a snarling V-12 engine and pugnacious good looks, the 333 SP—the first factory-built Ferrari prototype in a generation—won in its American debut at Road Atlanta in 1994 and dominated the sprint races for the rest of the season. But it didn't get a chance to contest any of the endurance classics until the next year. In the 1995 edition of the Rolex 24 at Daytona, two 333 SPs led until breaking halfway through the race. The following month, Ferraris qualified 1-3-4-6 for the 12-hour contest at Sebring, and all four cars led at some point during one of the wildest races in Sebring's history. Fat Florida raindrops started pelting the circuit an hour after the start, and at 6:20 p.m., a deluge of biblical proportions prompted a red flag that halted the race for seventy-five minutes. After tying the record for lead changes (twenty-three), the race came down to a long, punishing battle between a Chevy-powered Spice and the Scandia Motorsports 333 SP shared by team owner Andy Evans, Fermín Vélez, and Eric van de Poele. Although the Spice never managed to pass the Ferrari, it hounded the 333 SP for the rest of the night; with less than an hour to go, Vélez's lead dwindled to a mere ten seconds. But a late pit stop by the Spice gave Vélez enough breathing room to make a precautionary splash-and-go stop for fuel ten minutes before taking the checkered flag with 86.59 seconds in hand. This was Ferrari's first win at Sebring since 1972, when Mario Andretti and Jacky Ickx co-drove a 312 PB to victory. The 333 SP co-driven by Momo founder Gianpiero Moretti, who'd persuaded Ferrari to build the car, failed to finish.

RACES TO REMEMBER
1996 SPANISH GRAND PRIX

FUN FACT

Conditions were so tricky in Spain that Hill—who would win the championship that year—spun three times in eleven laps. "I am just relieved to be in one piece," he said after clouting the pit wall.

HISTORICAL TIDBIT

Here's how the great Stirling Moss—a frequent critic of Schumacher's—assessed the Grand Prix: "That was not a race. It was a demonstration of brilliance."

KEY PERSON

While Schumacher eked out three wins and finished third in the championship in 1996, Irvine managed just a single podium and DNFed in eight consecutive races.

At the Circuit de Barcelona-Catalunya in 1996, Michael Schumacher notched his first win for Ferrari while making the rest of the Formula 1 field look positively second rate. Schumi's maiden season for the team hadn't started strongly. The F310 was ungainly, unloved, and unreliable. In the first six races of the year, Schumacher had managed three podiums along with three DNFs, but he never looked like a winner. In Spain, he qualified third, nearly a second behind the pole-sitting Williams of Damon Hill. On race day, rain was falling so hard that there was talk about starting the race behind the safety car. When the lights went out, Schumacher nearly stalled his engine. By the time he slithered away from the grid, he'd dropped to sixth—a serious setback considering the appalling conditions. But Hill and Ferrari teammate Eddie Irvine spun early, and Schumacher quickly blew by Gerhard Berger's Benetton to climb up to third. On Lap 9, despite being nearly blinded by tire spray, he outbraked Jean Alesi's second Benetton in the second-gear Seat corner. Three laps later, he sliced past Jacques Villeneuve's Williams in the same spot to take the lead. Then, he walked on water. Within five laps, Schumacher was 14.9 seconds ahead of the field. Along the way, he set a fastest lap that was 2.2 seconds quicker than anybody else would go for the rest of the afternoon. Despite dropping a cylinder and losing power, Schumacher finished 45.302 seconds ahead of Alesi and lapped the field up to third place. A *regenmeister*—rainmaster—was born, and Ferrari had found the prophet who would lead the Scuderia to the Promised Land in years to come.

RACES TO REMEMBER
2015 MALAYSIAN GRAND PRIX

A shock, the pundits said. "Vettel stuns the Formula 1 world," the headlines read. Sebastian Vettel had won four consecutive World Championships before joining Scuderia Ferrari in 2015. But he was coming off a disastrous year when he'd been decisively outperformed by his teammate, and Ferrari was suffering through a humiliating thirty-four-race losing streak. Meanwhile, Mercedes had won sixteen of nineteen races in 2014 and opened the 2015 season with a 1-2 beatdown in Australia. Two weeks later, in Malaysia, everybody expected more of the same. Vettel managed to split the Mercedes drivers during qualifying only because of rain, and he barely held off Nico Rosberg at the start while pole-sitter Lewis Hamilton consolidated the lead. But an early safety-car period turned the race upside down. Defying conventional wisdom, Ferrari kept Vettel on the track while the other front-runners pitted. Because the SF15-T suffered less tire degradation than its rivals in the intense heat—the track temperature at the Sepang International Circuit was 142 degrees—Vettel was able to complete the race on two stops while Mercedes had to pit three times. Better still, his Ferrari was every bit as fast as the Silver Arrows, prompting some petulant complaints over the radio from an aggravated Hamilton. The Ferrari team was jubilant after Vettel won by 8.569 seconds, and an emotional Vettel teared up on the victory podium. "I remember when the gate opened in Maranello, it was like a dream coming true," he said. "I remember the last time I was there was as a young kid, watching Michael [Schumacher] over the fence driving around in the Ferrari, and now I'm driving that very red car. It's incredible."

END OF THE ROAD FOR THE MILLE MIGLIA

Racing was a blood sport for most of its first century, and death was no stranger to Ferrari. The honor roll of drivers who were killed in the red cars from Maranello includes Alberto Ascari, Lorenzo Bandini, Peter Collins, Luigi Musso, Pedro Rodriguez, Wolfgang von Trips, and Gilles Villeneuve. A Ferrari was also at the center of one of the worst disasters in the sport's history, at the Mille Miglia in 1957. This was a road race spanning nearly one thousand miles from Brescia to Rome and back again on public highways and serpentine mountain passes. Ferrari brought four fearsome sports racers and entrusted one of them to Spanish marquis Alfonso "Fon" de Portago. Besides being an international playboy and a superlative athlete, de Portago was also a talented race driver. But he abhorred the Mille Miglia. "There are hundreds of corners in the Mille Miglia where one slip by the driver can kill fifty people," he said presciently. De Portago made a stir at the checkpoint in Rome when his paramour, movie starlet Linda Christian, burst out of the crowd to passionately kiss him while cameras clicked. (Photographs of the gesture were inevitably captioned "the kiss of death.") But with his American friend Ed Nelson serving as his navigator, de Portago was outclassed in a race that favored local knowledge. He was fifth at the final fuel stop in Bologna, where Enzo Ferrari subtly urged him to make up time on the furrow-straight stretch through the Po Valley. At 175 miles per hour, about thirty miles from the finish, a tire exploded (or a half-shaft broke). The Ferrari spun, severed a telephone pole, flew 150 yards through the air, and cannoned into a ditch. De Portago and Nelson were killed instantly. Ten spectators, including five children, also died. White-haired Piero Taruffi won the race in a Ferrari 315 S. He immediately retired, and the Mille Miglia was never run again.

GLOSSARY

ACO: The Automobile Club de l'Ouest sanctions only one major race a year, but it's a doozy—the 24 Hours of Le Mans. Proudly French and inscrutably idiosyncratic, the ACO has ensured that Le Mans has remained one of racing's crown jewels since 1923.

FIA: The Fédération Internationale de l'Automobile is the largest and most influential sanctioning body in motorsports. Many countries have their own organizations, such as NASCAR, IndyCar, and IMSA in the United States. But the FIA oversees a host of international series, including Formula 1, the World Endurance Championship, and the World Rally Championship. It also promulgates safety standards that have been adopted worldwide.

FORMULA 1: A formula is a set of motorsports regulations governing everything from engine size to tire width. The FIA created Formula 1—the premier class of single-seat, open-wheel racing—in 1950. The first race was the British Grand Prix at Silverstone. Ferrari didn't start competing in F1 until the second race, at Monaco. And it didn't win an F1 race until the second British Grand Prix, again at Silverstone, in 1951.

PRIVATEERS: Current F1 rules require entrants to build the cars they race. But in years past, there were a handful of privateers who bought cars from other teams. The only privately entered Ferrari to win a World Championship event was the 156 driven by Giancarlo Baghetti in the French Grand Prix in 1961. Privateers are much more common in sports car racing. Both the first (in 1949) and the last (in 1965) Ferraris to win at Le Mans were private entries.

TEAM ORDERS: In the early days of racing, there was no prohibition against team orders. In fact, they were expected, and they played a critical role in how championships were won—and lost—in the 1950s and 1960s. The 1980s and 1990s, by contrast, were full of races where drivers flouted team orders and instead fought with each other. After Ferrari's Michael Schumacher and Rubens Barrichello swapped positions twice in 2002, the FIA banned team orders. But the new rules were impossible to enforce, and they were quietly dropped several years later.

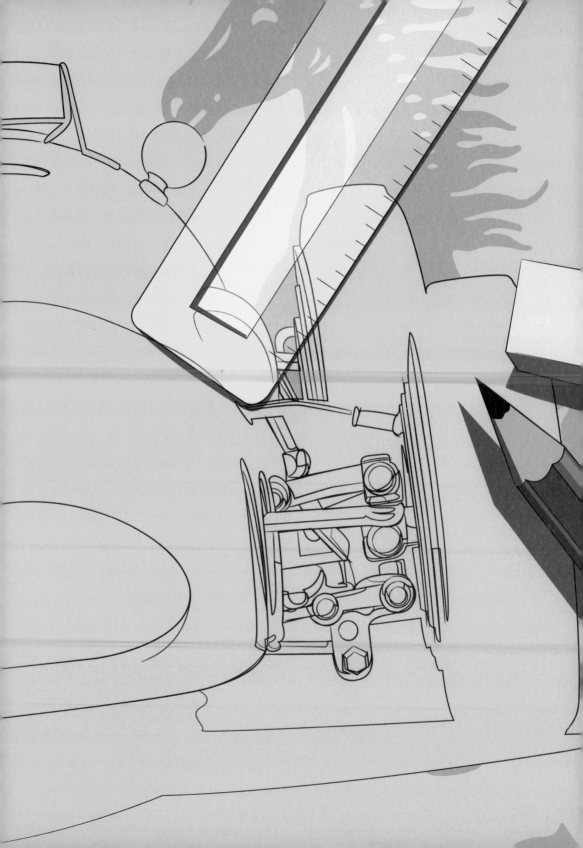

TEAM PLAYERS

FUN FACT

FUN FACT

The Ferrari 166 MM was Anderloni's first project after the death of his father, who'd founded Touring. "If people had not liked the design," he said, "it would have been the end of the company."

HISTORICAL TIDBIT

Race cars aside, Scaglietti's best-known project was transforming a savage 375 MM roadster into an elegant coupe for Roberto Rossellini, the film director, and his movie star wife, Ingrid Bergman—supposedly to prevent her hair from getting mussed.

KEY PERSON

Despite all of Michelotti's work with Ferrari, he reached a far wider audience designing sports cars for Triumph and, later, the British Leyland National bus.

Ferraris are all about performance and panache. The performance has always come straight from the factory, originally in the form of the signature V-12 engine. But the panache came not from Maranello but from Ferrari's chosen *carrozzerie*, which not only built the bodies but also designed them as well. So, let's give a shout out to Carlo Felice Bianchi Anderloni of Touring, whose egg-crate grille and impeccably proportioned 166 MM Barchetta created the "face" that launched countless Ferrari dealerships. Over the years, many of the most celebrated designers in Italy applied their genius to the red cars of Maranello. Giovanni Michelotti penned nearly a dozen early models for Vignale (and, a decade later, the one-off N.A.R.T. Spider for Luigi Chinetti), Giovanni Savonuzzi created the flamboyant tailfins for Ghia's 410 Superamerica, and even Marcello Gandini, best known for his Lamborghinis, has a Ferrari to his credit (the Dino 308 GT4 by Bertone). Although Sergio Scaglietti rarely styled street cars, he shaped a long line of iconic race cars ranging from the pontoon-fender Testa Rossa to the 250 GTO. Still, no coachbuilder was more closely associated with the marque than Pininfarina, which served as Ferrari's house designer. In the early days, it was Sergio Pininfarina himself. Later, the prolific Leonardo Fioravanti was the chief stylist behind a remarkable run of Ferraris—the Daytona, the Dino 206, the 308, the 288 GTO, the Berlinetta Boxer, the Testarossa, and the F40. Eventually, Ferrari created its own Styling Centre, and in 2013, the LaFerrari became the first Ferrari since 1973 without a trace of Pininfarina DNA. Contemporary Ferraris increasingly reflect the vision of current design chief Flavio Manzoni, and the Styling Centre's work has been showcased in the GTC4Lusso, 488 GTB, and 812 Superfast.

TEAM PLAYERS
COACHBUILDERS

Although Touring was responsible for the most famous of the 166 MMs, the Mille Miglia–winning Ferrari—which gave the MM its name—was the work of Carrozzeria Allemano. Despite the car's success, Allemano never built another Ferrari.

HISTORICAL TIDBIT

Ironically, the 612 Scaglietti was designed not by Scaglietti but by Pininfarina. Actually, this was standard practice for decades—Pininfarina styling the bodies and Scaglietti building them.

KEY PERSON

Sergio Scaglietti, the founder of his eponymous carrozzeria, was a member of Enzo Ferrari's so-called *amici del Sabato*, "Saturday friends," essentially Enzo's inner circle. He was one of the handful of old-timers invited to Ferrari's funeral.

Every country has its share of fine coachbuilders. But the *carrozzerie* of Italy have always been the first among equals, thanks to a seemingly endless supply of artisans who bring a special talent and passion to the industrial art of shaping metal. Without these shops, there would have been no Ferrari, which had neither the capacity nor the inclination to build bodies during its formative years. (As Enzo once said, "I build engines and attach wheels to them.") During the 1950s, cars were fashioned not only by hand but also by eye, and it was the job of the panel beaters to transform sketches into the rolling sculptures so prized by modern collectors. In the early days, Ferrari didn't play favorites, and he was happy to let the leading coach-builders compete for his favor. Carrozzeria Touring was responsible for the Barchetta that put Ferrari on the international map. But other 166 MMs were built by Vignale, Pinin Farina, and Zagato, while bodies for the 166 Inter, which was the marginally more user-friendly version of the MM, came from Touring, Vignale, Bertone, Ghia, and Stabilimenti Farina. Eventually, the list of Ferrari-sanctioned shops grew to include Fantuzzi, Drogo, Boano, and Ellena. But by the late 1950s, Ferrari's go-to coachbuilder was Carrozzeria Scaglietti. Although Scaglietti started on the race car side of the business, it was building virtually all the road cars as well by the 1960s. The company was bought by Fiat in 1969 and later subsumed into Ferrari, which continues to use the old Scaglietti plant to this day. The ties between the companies were enshrined in the name of the 2+2 grand touring Ferrari released in 2004—the 612 Scaglietti.

TEAM PLAYERS
ENGINEERS

When Enzo Ferrari went into business for himself, he poached so many engineers from his former employer that his company could have been called Alfa Romeo South. The most crucial hire was Gioacchino Colombo, formerly the right-hand man of illustrious Alfetta designer Vittorio Jano, who himself later joined Ferrari and conceptualized the Dino line of V-6s. Colombo created Ferrari's first engine, the 1.5-liter supercharged V-12, whose basic architecture survived into the late 1960s. But by that time, Colombo had been replaced by his former underling, Aurelio Lampredi. Unlike most of his colleagues, Lampredi came from the aviation world, and he thought outside the box: He was the engineer who designed the unsupercharged V-12 that dethroned the Alfetta. Plus, he was responsible for the four-cylinder engine that made the Ferrari 500 virtually invincible. But Enzo was a tough man to work for, and Lampredi's days were numbered. By the end of the 1950s, the stars of the engineering department were two more Alfa refugees—Carlo Chiti, best known for the shark-nose Formula 1 car, and Giotto Bizzarrini, the father of the 250 GTO. But they, too, had Enzo issues, and they were fired in the great Purge of 1961. This left twenty-seven-year-old Mauro Forghieri in charge. He and engine man Franco Rocchi responded by developing the championship-winning V-8 of 1964 and the legendary flat-twelves of the 1970s. After that, F1 chassis gurus took precedence over engine designers—and they were Brits, no less! First came Harvey Postlethwaite (the twin-turbo 126 C2) and John Barnard (the semi-automatic 640). But Ferrari reached its motorsport apogee when Englishman Ross Brawn and South African Rory Byrne joined Michael Schumacher at the resurgent scuderia in 1997. The result? Six straight constructors' titles. Enzo would have approved.

Cars don't sell themselves, not even Ferraris. While Enzo Ferrari was fixated on racing, he knew he had to sell street cars to support his scuderia. Few men played a more important role in getting the marque established than Girolamo Gardini, who managed the sales department until being sacked in the Purge of 1961. Then, as now, foreign sales were critical to Ferrari. No market was larger or more important than the United States, thanks in large part to the redoubtable and essential Luigi Chinetti, who won Le Mans twice in Alfa Romeos before the war and again in a Ferrari in 1949. Chinetti told Enzo in 1946 that he ought to quit the machine-tool business and start selling cars. When Enzo asked who would buy them, Chinetti replied: "I will buy twenty, twenty-five cars." Decades later, Chinetti told journalist Denise McCluggage, "Can you imagine? I have no money and I say I will buy twenty-five cars." After briefly running a dealership in Paris, Chinetti became the first—and for many years the only—Ferrari importer in the States, and he later formed the North American Racing Team (N.A.R.T.). Chinetti's counterpart in the United Kingdom was Colonel Ronnie Hoare, a retired Royal Artillery officer who became the country's Ferrari distributor in 1960. Although he sold only four cars that first year, Hoare was soon successful enough to move the dealership—Maranello Concessionaires—into a historic Art Deco building and form a successful race team of his own. Another prominent dealer/racer was Belgian Jacques Swaters, the first Ferrari distributor in Europe. In 1965, a 250 LM in the yellow colors made famous by Swaters's Ecurie Francorchamps finished second at Le Mans—to another 250 LM entered by N.A.R.T.

TEAM PLAYERS
EXECUTIVES

Enzo Ferrari ruled his company as an absolute monarch. He had a retinue of loyal retainers, chief among them Franco Gozzi, who was both his confidante and his press agent. But even Ferrari was sometimes unable to corral his tempestuous wife, the former Laura Garello. Although she was muted in public, her meddling at the factory prompted corporate discontent that caused the Purge of 1961. Despite his long marriage, Ferrari maintained a mistress, Lina Lardi, most of his life. With Lardi, he had a son, Piero, who went to work at the factory—studiously avoiding Ferrari's wife—in 1965. When Laura died in 1978, he was formally acknowledged as Piero Ferrari. After Enzo died in 1988, Piero inherited 10 percent of the company, which he retains to this day. As vice chairman, he's played a particularly important role on the supercar side of the business. Enzo's death briefly left a vacuum that was soon filled by the urbane Luca di Montezemolo, who'd already done a brief tour of duty reviving the Formula 1 team in the mid-1970s. Montezemolo instituted new policies and infused the company with fresh blood. Frenchman Jean Todt, who had been running Peugeot's Le Mans program, was brought in to oversee the F1 team, while Amedeo Felisa became one of the few engineers elevated to serve as CEO of a major automaker. But Montezemolo's brash style clashed with the bottom-line mentality of Fiat Chrysler Automobiles chief Sergio Marchionne. In 2014, Montezemolo left Ferrari after being told, "No one is indispensable." Two years later, Ferrari was spun off from FCA, and Felisa retired shortly thereafter. Today, Marchionne rules Ferrari almost as autocratically as Enzo did three decades earlier.

PISTA DI FIORANO

In the 1950s and 1960s, residents of Modena were accustomed to seeing Ferrari race cars trundling noisily through town with their exhausts popping. Their destination was the airport on the western edge of the city, where a simple race circuit had been fashioned out of runways and taxiways. But the track wasn't useful for testing as cars got faster and more sophisticated. So, in 1972, Enzo Ferrari built a private circuit on a tract of land he'd bought in Fiorano, about a half-mile from the factory in Maranello. Over the years, it's been lengthened and upgraded with embedded sensors (to log data) and sprinklers (to simulate wet conditions). The Pista di Fiorano now consists of twelve turns laid out in a rough figure eight measuring 1.862 miles. Each section of the track is designed to evaluate a specific attribute of the car—handling, braking, aerodynamics, etc. Although the circuit isn't meant for racing, the stopwatch is always running. The fastest road car around Fiorano is the LaFerrari, which clocked a lap time of 1:19.7. But the all-time record is held by Michael Schumacher in an F2004 Formula 1 car at 55.999 seconds, which translates into 119.70 miles per hour. Enzo loved watching F1 testing from his lair in a converted farmhouse with whitewashed walls and red-timber shutters. (His office has been preserved down to the black-and-white television he used for watching F1 broadcasts.) And during his tenure running the company, Luca di Montezemolo routinely landed on the helipad to hold court for journalists and visiting VIPs. The circuit isn't open to the public. But anybody can stand outside the chainlink fence lining the track and, with any luck, glimpse longtime test driver Dario Benuzzi sliding a Ferrari around a corner.

GLOSSARY

FCA: Chrysler filed for bankruptcy in 2009. Later that year, the company emerged from Chapter 11 by forming a global alliance that gave Fiat a 20 percent stake—later expanded to 35 percent—in the company. In 2014, a new conglomerate called Fiat Chrysler Automobiles was formed. Its name notwithstanding, FCA also incorporated Ferrari, Alfa Romeo, Maserati, and Lancia. Two years later, Ferrari was spun off from FCA as a separate, freestanding entity.

FIAT: The Società Anonima Fabbrica Italiana di Automobili Torino—originally known as F.I.A.T., with periods between the letters—was formed in 1899. The following year, it built twenty-four cars, and Giovanni Agnelli became managing director in 1902. Fiat was a powerhouse in international motorsports before withdrawing from racing in 1927 to focus on selling prodigious numbers of road cars. In 1969, the company—still run by the Agnelli family—bought a 40 percent stake in Ferrari.

ONE-OFF: The term "one-off" generally refers to a car featuring a custom body on a production chassis. Though never commonplace, one-offs were much easier to build in the 1950s and 1960s, when body-on-frame construction was the standard. Some one-off Ferraris became famous (the 410 Superfast I). Some became infamous (the doorless, roofless 365 GTC/4 Beach Car). Some became a little bit of both (e.g., the Breadvan). In recent years, Ferrari has brought the customization philosophy in-house by creating sanctioned one-offs through its Styling Centre.

THE PURGE: Also known as the Palace Revolt, the Purge referred to the mass firing of Ferrari department heads at the end of 1961. Although the exact cause of the episode is unclear, it was evidently related to disgruntlement with the role played by Enzo Ferrari's wife, Laura. By the time the dust settled, much of Ferrari's senior leadership had been axed. The victims included technical director Carlo Chiti, chief development engineer Giotto Bizzarrini, race team manager Romolo Tavoni, and sales director Girolamo Gardini. Several of the refugees went to work for a short-lived marque called ATS—for Automobili Turismo e Sport—and collaborated on an unsuccessful Formula 1 car driven unhappily by Phil Hill in 1963.

INDEX

FERRARI TIMELINE

1898 Enzo Ferrari is born in Modena, the son of the owner of a small machine shop.

1933 Scuderia Ferrari takes control of Alfa Romeo's Grand Prix
and sports car racing programs.

1947 The first Ferrari, the 125 S, breaks in its debut race in Piacenza.
But it wins in Rome two weeks later.

1951 The Ferrari 375 scores the marque's first World Championship
Grand Prix victory, at Silverstone.

1959 The much-loved 250 GT SWB debuts. Is it a street car or is it a race car? Both.

1969 Fiat buys 40 percent of Ferrari and takes control of production cars
while Enzo oversees the race program.

1973 The 365 GT4 BB, better known as the Berlinetta Boxer,
becomes the first production car with a flat-twelve engine.

1982 Gilles Villeneuve, the most spectacular driver of his era,
is killed during qualifying for the Belgian Grand Prix.

1988 Enzo Ferrari dies at age 90. The F40, the last car he personally approved,
debuted the year before.

2000 Michael Schumacher wins his first World Championship for Ferrari.
He then wins the next four titles consecutively.

2014 A 250 GTO sells at auction for $38.1 million, the highest price ever paid
publicly for an automobile.

2016 Ferrari N.V. is spun off from Fiat Chrysler Automobiles to become
a separate, freestanding company.